GIVING FULL MEASURE TO COUNTERMEASURES

Addressing Problems in the DoD Program to Develop Medical Countermeasures Against Biological Warfare Agents

Committee on Accelerating the Research, Development, and Acquisition of Medical Countermeasures Against Biological Warfare Agents

Medical Follow-up Agency
and
Board on Life Sciences

Lois M. Joellenbeck, Jane S. Durch, and Leslie Z. Benet, *Editors*

INSTITUTE OF MEDICINE *AND*
NATIONAL RESEARCH COUNCIL
OF THE NATIONAL ACADEMIES

THE NATIONAL ACADEMIES PRESS
Washington, D.C.
www.nap.edu

THE NATIONAL ACADEMIES PRESS 500 Fifth Street, N.W. Washington, DC 20001

NOTICE: The project that is the subject of this report was approved by the Governing Board of the National Research Council, whose members are drawn from the councils of the National Academy of Sciences, the National Academy of Engineering, and the Institute of Medicine. The members of the committee responsible for the report were chosen for their special competences and with regard for appropriate balance.

Support for this project was provided by Contract No. DAMD17-02-C-0099 between the National Academy of Sciences and the U.S. Army. The views presented in this report are those of the Institute of Medicine and National Research Council Committee on Accelerating the Research, Development, and Acquisition of Medical Countermeasures Against Biological Warfare Agents and are not necessarily those of the funding agencies.

Library of Congress Cataloging-in-Publication Data

Giving full measure to countermeasures : addressing problems in the DOD program to develop medical countermeasures against biological warfare agents / Committee on Accelerating the Research, Development, and Acquisition of Medical Countermeasures against Biological Warfare Agents, Medical Follow-up Agency and Board on Life Sciences ; Lois M. Joellenbeck, Jane S. Durch, and Leslie Z. Benet, editors.

p. ; cm.

Includes bibliographical references.

ISBN 0-309-09153-5 (pbk.)

1. Biological warfare—Health aspects—United States. 2. Biological warfare—United States—Safety measures. 3. Biological weapons—Health aspects—United States. 4. Biological weapons—United States—Safety measures. 5. Medicine, Preventive—United States. 6. Antitoxins—United States. 7. Vaccines—United States.

[DNLM: 1. United States. Dept. of Defense. 2. Biological Warfare—prevention & control. 3. Drug Approval. 4. Antitoxins. 5. Military Personnel. 6. United States Government Agencies. 7. Vaccines.] I. Joellenbeck, Lois M. (Lois Mary), 1963- II. Durch, Jane. III. Benet, Leslie Z. IV. Committee on Accelerating the Research, Development, and Acquisition of Medical Countermeasures against Biological Warfare Agents.

UG447.8.G58 2004

358'.38—dc22

2004005890

Additional copies of this report are available from the National Academies Press, 500 Fifth Street, N.W., Lockbox 285, Washington, DC 20055; (800) 624-6242 or (202) 334-3313 (in the Washington metropolitan area); Internet, http://www.nap.edu.

For more information about the Institute of Medicine, visit the IOM home page at: **www.iom.edu.**

Printed in the United States of America.

Cover credit: The upper cover photograph is from a U.S. Department of Defense news photograph. [Online]. Available: http://www.defenselink.mil/photos/May1997/970422-F-7719S-002.html.

THE NATIONAL ACADEMIES

Advisers to the Nation on Science, Engineering, and Medicine

The **National Academy of Sciences** is a private, nonprofit, self-perpetuating society of distinguished scholars engaged in scientific and engineering research, dedicated to the furtherance of science and technology and to their use for the general welfare. Upon the authority of the charter granted to it by the Congress in 1863, the Academy has a mandate that requires it to advise the federal government on scientific and technical matters. Dr. Bruce M. Alberts is president of the National Academy of Sciences.

The **National Academy of Engineering** was established in 1964, under the charter of the National Academy of Sciences, as a parallel organization of outstanding engineers. It is autonomous in its administration and in the selection of its members, sharing with the National Academy of Sciences the responsibility for advising the federal government. The National Academy of Engineering also sponsors engineering programs aimed at meeting national needs, encourages education and research, and recognizes the superior achievements of engineers. Dr. Wm. A. Wulf is president of the National Academy of Engineering.

The **Institute of Medicine** was established in 1970 by the National Academy of Sciences to secure the services of eminent members of appropriate professions in the examination of policy matters pertaining to the health of the public. The Institute acts under the responsibility given to the National Academy of Sciences by its congressional charter to be an adviser to the federal government and, upon its own initiative, to identify issues of medical care, research, and education. Dr. Harvey V. Fineberg is president of the Institute of Medicine.

The **National Research Council** was organized by the National Academy of Sciences in 1916 to associate the broad community of science and technology with the Academy's purposes of furthering knowledge and advising the federal government. Functioning in accordance with general policies determined by the Academy, the Council has become the principal operating agency of both the National Academy of Sciences and the National Academy of Engineering in providing services to the government, the public, and the scientific and engineering communities. The Council is administered jointly by both Academies and the Institute of Medicine. Dr. Bruce M. Alberts and Dr. Wm. A. Wulf are chair and vice chair, respectively, of the National Research Council.

www.national-academies.org

COMMITTEE ON ACCELERATING THE RESEARCH, DEVELOPMENT, AND ACQUISITION OF MEDICAL COUNTERMEASURES AGAINST BIOLOGICAL WARFARE AGENTS

LESLIE Z. BENET (*Chair*), Professor, Department of Biopharmaceutical Sciences, School of Pharmacy, University of California, San Francisco
WALTER E. BRANDT, Senior Program Officer, Malaria Vaccine Initiative, Program for Appropriate Technology in Health
BARRY S. COLLER, Vice President for Medical Affairs, David Rockefeller Professor of Medicine, and Head of the Laboratory of Blood and Vascular Biology, The Rockefeller University
GLENNA M. CROOKS, President, Strategic Health Policy International, Inc.
R. GORDON DOUGLAS, JR., Scientific Consultant
JACQUES S. GANSLER, Professor and Roger C. Lipitz Chair, Center for Public Policy and Private Enterprise, School of Public Affairs, University of Maryland
ANTHONY L. ITTEILAG, Independent Consultant
DENNIS L. KASPER, Executive Dean for Academic Programs, William Ellery Channing Professor of Medicine, and Professor of Microbiology and Molecular Genetics, Harvard Medical School
STEVEN J. KELMAN, Albert J. Weatherhead III and Richard W. Weatherhead Professor of Public Management, John F. Kennedy School of Government, Harvard University
RICHARD F. KINGHAM, Partner, Covington and Burling
PETER M. PALESE, Professor and Chair, Department of Microbiology, Mount Sinai School of Medicine
PAUL D. PARKMAN, President, Parkman Associates
RONALD J. SALDARINI, Scientific Consultant
JANE E. SISK, Professor, Department of Health Policy, Mount Sinai School of Medicine
ELAINE I. TUOMANEN, Professor and Chair, Department of Infectious Diseases, St. Jude Children's Research Hospital
BENJAMIN J. WEIGLER, Director, Animal Health Resources, Division of Shared Resources, Fred Hutchinson Cancer Research Center
JANET WESTPHELING, Associate Professor, Department of Genetics, University of Georgia

Medical Follow-up Agency Board Liaison

TIMOTHY R. GERRITY, Director, Bioengineering Institute, Worcester Polytechnic Institute

Consultant

JAMES D. MARKS, Professor, Anesthesia and Pharmaceutical Chemistry, University of California, San Francisco

Project Staff

LOIS JOELLENBECK, Senior Program Officer, Medical Follow-up Agency
JANE DURCH, Senior Program Officer, Medical Follow-up Agency
KAREN KAZMERZAK, Research Associate, Medical Follow-up Agency (through July 2003)
PHILLIP BAILEY, Senior Project Assistant, Medical Follow-up Agency

Auxiliary Staff

RICHARD N. MILLER, Director, Medical Follow-up Agency (to November 2003)
RICK ERDTMANN, Director, Medical Follow-up Agency (since November 2003)
JENNIFER KUZMA, Senior Program Officer, Board on Life Sciences, Division on Earth and Life Studies (until January 2003)
FRANCES SHARPLES, Director, Board on Life Sciences, Division on Earth and Life Studies
REINE HOMAWOO, Senior Project Assistant, Medical Follow-up Agency
PAMELA RAMEY-MCCRAY, Administrative Assistant, Medical Follow-up Agency
ANDREA COHEN, Financial Associate, Institute of Medicine

Reviewers

This report has been reviewed in draft form by individuals chosen for their diverse perspectives and technical expertise, in accordance with procedures approved by the National Research Council's Report Review Committee. The purpose of this independent review is to provide candid and critical comments that will assist the institution in making its published report as sound as possible and to ensure that the report meets institutional standards for objectivity, evidence, and responsiveness to the study charge. The review comments and draft manuscript remain confidential to protect the integrity of the deliberative process. We wish to thank the following individuals for their review of this report:

JAMES L. BLANCHARD, Tulane Regional Primate Center
WILLIAM H. HABIG, Centocor, Inc.
ROBERT HELMS, The American Enterprise Institute
BRUCE KUHLIK, Pharmaceutical Research and Manufacturers of America
MYRON M. LEVINE, University of Maryland at Baltimore
JOHN S. PARKER, Science Applications International Corporation
HOWARD R. SIX, Aventis Pasteur (retired)
FRED THOMPSON, Willamette University
JOHN P. WHITE, Harvard University
GEORGE WHITESIDES, Harvard University

Although the reviewers listed above have provided many constructive comments and suggestions, they were not asked to endorse the conclusions or recommendations nor did they see the final draft of the report before its release. The review of this report was overseen by Alexander Flax, consultant, and Morton Swartz, Massachusetts General Hospital. Appointed by the National Research Council and the Institute of Medicine, they were responsible for making certain that an independent examination of this report was carried out in accordance with institutional procedures and that all review comments were carefully considered. Responsibility for the final content of this report rests entirely with the authoring committee and the institution.

Preface

The Gulf War of 1990–1991 renewed Cold War concerns that U.S. service members might be exposed to chemical or biological warfare agents on the battlefield. These concerns were reinforced after the war upon discovery of Iraqi stockpiles of weaponized biological and chemical agents.[1] In 2001, the distribution of *Bacillus anthracis* spores through the U.S. postal system renewed public awareness of the threats posed by biological agents.

At the time of the Gulf War, only one medical countermeasure approved by the Food and Drug Administration (FDA)—the vaccine against anthrax—was available to the Department of Defense (DoD) to protect troops against possible biological warfare agents. In 2003, despite congressional attention and good-faith efforts on the part of DoD scientists, no new vaccines against biowarfare agents are available to service members.[2]

In the National Defense Authorization Act for Fiscal Year 2002 (P.L. 107-107), Congress directed the Secretary of Defense to accelerate the

[1]United Nations Special Commission. 1999. UNSCOM: Chronology of Main Events. New York: United Nations. [Online]. Available: http://www.un.org/Depts/unscom/Chronology/chronologyframe.htm.

[2]Since 2000, the FDA has renewed the product license for existing supplies of smallpox vaccine and has approved labeling ciprofloxacin (Cipro), tetracyclines (including doxycycline), and penicillins for treatment of anthrax.

department's efforts to develop FDA-licensed medical countermeasures against biological warfare agents. In addition, the Secretary was directed to contract with the Institute of Medicine (IOM) and the National Research Council (NRC) for a study of the review and approval process for new medical countermeasures in order to identify new approaches to accelerate that process and to identify methods for ensuring that new countermeasures will be safe and effective. To carry out the study, IOM and NRC convened the Committee on Accelerating the Research, Development, and Acquisition of Medical Countermeasures Against Biological Warfare Agents.

The members of the committee bring to the study expertise in drug and vaccine acquisition in DoD; vaccine and drug research, development, testing, and evaluation in academia and the pharmaceutical and biotechnology industries; laboratory animal science; federal drug and vaccine regulatory policy; legal and economic issues in drug and vaccine development; and management processes in government and industry.

The committee has the following charge:

> The committee will examine DoD's biowarfare countermeasure drug and vaccine acquisition process. The acquisition process includes the early science and technology development (research and development program elements 6.1, 6.2, 6.3) and advanced development (program elements 6.4, 6.5) through the approval and licensure of products. The study will not examine production and procurement processes. The committee will identify factors that are impeding or slowing the acquisition processes and will recommend strategies or options for accelerating these processes.

Guided by discussion with DoD representatives and congressional staff at its first meeting, the committee interpreted this charge as calling for its work to focus on the manner in which DoD organizes and manages research, development, and acquisition of medical countermeasures, rather than on the details of specific scientific approaches. The medical products covered by the study include vaccines, antitoxins, chemoprophylactics, and chemotherapeutics. In keeping with its charge, the committee did not examine the acquisition of diagnostic products or other biodefense products, such as protective suits, decontamination equipment, or sensors for detection of biological agents, that are also being developed as essential components of DoD's Chemical and Biological Defense Program. The military services and combatant commands have an important role in these broader aspects of biodefense that were not a focus of this study.

Previous independent advisory committees requested by DoD released reports in 2000 and 2002[3] addressing aspects of the development and production of vaccines for military needs and providing important context for the present study. Because those and other reports have dealt in detail with some of the topics covered here, the committee chose in this report to focus attention on the opportunities it identified for improvement, rather than on extensive historical and background information. To that end, the report begins with the committee's major recommendations in Chapters 1 and 2, with additional, second-order recommendations in Chapter 3. Brief background material on the current DoD program to develop medical biowarfare countermeasures and the current status of biowarfare countermeasures is provided in Appendix A.

The committee met six times between December 2002 and July 2003. At five of those meetings, the committee met in public session for presentations from and discussions with invited speakers (see Appendix B for the agendas of the public sessions). Small subgroups of committee members also participated in a series of informal meetings with officials from the several DoD organizations with a role in the current Chemical and Biological Defense Program (see Figure 2-1 for an organizational chart), with congressional staff, and with representatives from DynPort Vaccine Company, the National Institute of Allergy and Infectious Diseases (NIAID) of the National Institutes of Health, the FDA, and the Department of Homeland Security.

An interim report was submitted to the sponsor in March 2003.

As chair, I am very grateful to my fellow committee members for the dedication and industry that they exhibited in addressing a very complex and difficult problem. They willingly and enthusiastically participated in the six formal meetings of the committee, and many were available for the 20 informal meetings with various governmental and other relevant contacts as described above. Finally, the entire committee participated in four conference calls as this report evolved. The committee was frequently presented with very contradictory opinions and recommendations, while the landscape continually changed as the administration's response to perceived bioterrorism threats and the progress of the war in Iraq brought different perspectives to our charge.

---

[3]Institute of Medicine. 2002. *Protecting Our Forces: Improving Vaccine Acquisition and Availability in the U.S. Military*. Lemon SM, Thaul S, Fisseha S, O'Maonaigh HC, eds. Washington, DC: National Academies Press. Top FH Jr., Dingerdissen JJ, Habig WH, Quinnan GV Jr., Wells RL. 2000. DoD Acquisition of Vaccine Production: Report to the Deputy Secretary of Defense by the Independent Panel of Experts. In DoD. 2001. *Report on Biological Warfare Defense Vaccine Research and Development Programs*. Washington, DC: Department of Defense. [Online]. Available: http://www.acq.osd.mil/cp/bwdvrdp-july01.pdf.

The committee is also in debt to numerous officials from DoD, congressional staff, senior leadership staff from DynPort Vaccine Company, NIAID, FDA, and the Department of Homeland Security who assisted the committee in its fact finding, and in particular our study contact with DoD, Dr. Carol Linden. These individuals, as noted in Appendix C, graciously appeared before the committee at our formal meetings, sometimes two or three times, to make presentations. In addition many of them often attended the committee meetings when they were not making a formal presentation, and they were always willing to be available to meet informally with subsets of the committee and to respond to questions from the staff. We could not have completed our work in a timely manner without the cooperation of these officials.

The committee and I are extremely grateful to the outstanding IOM staff that tirelessly and with unselfish dedication helped to prepare this report in a timely manner. We are particularly appreciative of our two Senior Program Officers Lois Joellenbeck and Jane Durch, and the committee certifies that they are jewels, whose brilliance and worth shone brightly in enlightening and assisting committee members in evaluating the many controversial and technical issues related to the study. The committee also acknowledges and appreciated the work of Research Associate Karen Kazmerzak, and Senior Project Assistants Phillip Bailey and Reine Homawoo, who were unflagging in their effective responses to committee needs. This level of commendation speaks well of Richard N. Miller, the director of the Medical Follow-up Agency, who has assembled such a responsive staff, and the committee thanks him for his support and frequent inobtrusive input into its deliberations. We are grateful as well to Bronwyn Schrecker, Clyde Behney, Janice Mehler, Jennifer Bitticks, Jennifer Otten, and Andrea Cohen, the IOM and NRC staff who participated in the report review, preproduction, dissemination, and financial management for the report.

Leslie Z. Benet, Ph.D.
Chair

Contents

Figures, Tables, and Boxes

FIGURES

TABLES

BOXES

Abbreviations and Acronyms

AAE	Army Acquisition Executive (Secretary of the Army)
ASD(HA)	Assistant Secretary of Defense for Health Affairs
ATSD(NCB)	Assistant to the Secretary of Defense for Nuclear and Chemical and Biological Defense Programs
BoT	botulinum toxoid vaccine
BSL	biosafety level
CBD	Chemical and Biological Defense Directorate
CBDP	Chemical and Biological Defense Program
CBMS	Chemical Biological Medical Systems
CDC	Centers for Disease Control and Prevention
C.F.R.	Code of Federal Regulations
cGMP	current Good Manufacturing Practice
CJCS	Chairman of the Joint Chiefs of Staff
CRADA	Cooperative Research and Development Agreement
DARPA	Defense Advanced Research Projects Agency
DATSD(CBD)	Deputy Assistant to the Secretary of Defense for Chemical and Biological Defense
DHHS	Department of Health and Human Services
DHS	Department of Homeland Security
DoD	Department of Defense
DTRA	Defense Threat Reduction Agency
DUS&T	Dual-Use Science and Technology

DVC DynPort Vaccine Company

FDA Food and Drug Administration
FY fiscal year

GAO General Accounting Office
GMP Good Manufacturing Practice

IND Investigational New Drug
IOM Institute of Medicine

JPEO Joint Program Executive Office
JPEO-CBD Joint Program Executive Office for Chemical and Biological Defense
JRO Joint Requirements Office
JRO-CBRN Joint Requirements Office for Chemical, Biological, Radiological, and Nuclear Defense
JVAP Joint Vaccine Acquisition Program

MBDRP Medical Biological Defense Research Program
MCBDRP Medical Chemical and Biological Defense Research Program
MCDRP Medical Chemical Defense Research Program
MIDRP Military Infectious Diseases Research Program
MITS Medical Identification and Treatment Systems
MRMC Medical Research and Materiel Command

NDA New Drug Application
NIAID National Institute of Allergy and Infectious Diseases
NIH National Institutes of Health
NRC National Research Council

OTA Office of Technology Assessment

PDUFA Prescription Drug User Fee Act
P.L. Public Law

SAFETY Act Support of Anti-terrorism by Fostering Effective Technologies Act of 2002
SBIR Small Business Innovation Research
SECDEF Secretary of Defense
STTR Small Business Technology Transfer

USAMMDA	U.S. Army Medical Materiel Development Activity
USAMRIID	U.S. Army Medical Research Institute of Infectious Diseases
USAMRMC	U.S. Army Medical Research and Materiel Command
U.S.C.	United States Code
USD	Under Secretary of Defense
USD(AT&L)	Under Secretary of Defense for Acquisition, Technology, and Logistics
USD(P&R)	Under Secretary of Defense for Personnel and Readiness
WRAIR	Walter Reed Army Institute of Research

Executive Summary

The biodefense efforts of the Department of Defense (DoD) are poorly organized to develop and license vaccines, therapeutic drugs, and antitoxins to protect members of the armed forces against biological warfare agents.

These efforts are characterized by fragmentation of responsibility and authority, changing strategies that have resulted in lost time and expertise, and a lack of financial commitment commensurate with the requirements of program goals. These factors, together with special regulatory challenges for obtaining Food and Drug Administration (FDA) approval of biowarfare countermeasures,[1] mean that since the Gulf War of 1990–1991 DoD has gained no new vaccines and only a few drugs as medical biodefense countermeasures.

This serious situation exists despite declarations that biological warfare poses a significant threat to the safety and effectiveness of the nation's armed forces (Bush, 2002; Cohen, 1997; Defense Science Board, 2001, 2002; Perry, 1996; U.S. Congress, House Armed Services Committee, 1993), the recent vaccination of large numbers of military personnel against anthrax and smallpox, a DoD commitment to acquire vaccines against all vali-

[1]It was not possible to license new vaccines or drugs against biological warfare agents until July 2002, when the FDA's "Animal Efficacy Rule" became effective (FDA, 2002). The Animal Efficacy Rule allows the use of efficacy data from animal studies when tests of efficacy in humans are not ethical or feasible, as is generally the case with medical biodefense countermeasures.

dated biological warfare threats (DoD, 1993), and concerns about new bioengineered microbial threats.

The development and licensure of new vaccines and drugs is a difficult, expensive, and time-consuming process. Moreover, biodefense products pose special scientific, regulatory, and ethical challenges because it is generally unacceptable to expose humans to biowarfare agents to establish the efficacy of those products. Accelerating the development and licensure of such products will require strong and creative scientific leadership and a sustained commitment of adequate financial and other resources.

This report presents the findings and recommendations of a committee of the Institute of Medicine (IOM) and the National Research Council (NRC) convened to conduct a study mandated by Congress. The charge to the committee is to recommend strategies for accelerating the DoD research, development, and licensure processes for new medical biodefense countermeasures.[2,3] Based on the study charge to address strategies for accelerating these processes, the committee focused its attention on the organization and management of these processes, rather than on details of specific scientific approaches.

The committee was not asked to assess the nature or extent of any biological warfare threat or to compare the value to DoD of developing medical countermeasures against biological warfare agents relative to the pursuit of its other obligations. The committee viewed its task as resting on the premise that biological weapons pose a genuine threat to the health of military personnel, and therefore additional FDA-licensed medical countermeasures are urgently needed.

THE CURRENT CONTEXT

Scientific and technological developments are expanding the range of potential biological threats as well as opening pathways to new countermeasures. Meanwhile the need continues for work on countermeasures currently in development pipelines to protect against more familiar biological threats. For decades, DoD carried out the only significant effort in medical biodefense countermeasure development. Now, however, a substantial biodefense research effort is under way within the Department of

[2]The study was called for in the National Defense Authorization Act for Fiscal Year 2002 (P.L. 107-107). The study charge, the expertise of the committee, and the committee's approach to the study are discussed in the Preface.

[3]In this report, the term "licensed" is used to connote approval by the FDA of either a Biologics License Application or New Drug Application.

Health and Human Services (DHHS), with funding of $1.7 billion allocated to the National Institutes of Health (NIH) for fiscal year (FY) 2003 (and $1.6 billion requested for FY 2004) (DHHS, no date). In addition, a proposal for "Project BioShield" aims to create incentives for the pharmaceutical industry to manufacture and license medical countermeasures by making up to $6 billion dollars available over the next 10 years to purchase those products for a national stockpile (White House, 2003). In contrast, DoD's funding for its research and development program for medical countermeasures against at least 10 biological agents amounts to only $267 million for FY 2003 (DoD, 2003; Evans, 2003).[4]

The upsurge in funding and effort aimed at protecting the civilian population against bioterrorism will undoubtedly result in the development of new technologies and products that can also aid in protecting military personnel against the risks of biological warfare. However, the perceived risks posed by these agents can be different in the two settings, and there are important differences between the planning for protection against bioterrorism and biological warfare (DoD, 1993, 2000; Fauci, 2003; Linden, 2002). The military has considered vaccination to be the primary medical strategy for battlefield protection of a defined and relatively small population. Mass vaccination of the civilian population against a range of potential biological threats is less appropriate and much less feasible. For DoD, the aim is to protect service members in a manner that allows them to maintain their combat effectiveness and limits the need for medical personnel and equipment to treat casualties. Having the capacity for rapid diagnosis and postexposure treatment is also essential for DoD, but it is less desirable as a primary strategy for protecting troops on the battlefield than it is for responding to a bioterrorism event in the civilian population.

The challenges in developing new vaccines and drugs include the cost of the process and the substantial risk of failure. The congressional Office of Technology Assessment (OTA) in 1993 estimated the average cost of bringing a new drug to licensure to be $237 million (1990 dollars) (U.S. Congress, OTA, 1993). More recent estimates have ranged from $110 million to $802 million (2000 dollars) (DiMasi et al., 2003; Public Citizen, 2001). As few as one candidate in 5,000 reaches clinical testing, and only 20 percent of candidates that begin clinical testing reach licensure (FDA, 1999;

---

[4]The DoD funding for FY 2003 includes program elements for medical biological defense in the Chemical and Biological Defense Program (under budget activities 6.1–6.5) and biomedical components of the biological warfare defense program of the Defense Advanced Research Projects Agency (DARPA). The funding totals for medical biodefense exclude costs presently covered in accounts of the military services for salaries and benefits for military personnel and for operating certain facilities.

NVAC, 1999; PhRMA, 2000; Struck, 1996). Such estimates are based primarily on data for new drugs; equivalent estimates for vaccines and other biologics are rarely presented (IOM, 2003).

The process is also time consuming, with an industry estimate of 7 to 12 years for vaccine development, but with past experience showing that successful completion of clinical testing alone can take as long as 20 years (Grant, 2003; NVAC, 1999). New techniques are likely to speed the discovery of some candidate countermeasures, but are unlikely to accelerate some of the most time-consuming parts of the product development process, including the crucial assessment of product safety in human volunteers and efficacy based on animal data under new FDA regulatory guidelines (the "Animal Efficacy Rule") (FDA, 2002). The use of animal-based efficacy testing for products intended to protect against potentially lethal pathogens for which efficacy studies in humans are not feasible or ethical is still a new process and likely to require considerable time and effort to become regularized.

PROBLEMS WITH THE DOD EFFORT

The committee sees dismal prospects for successful results (and no prospects for faster results) from the current efforts by DoD's Chemical and Biological Defense Program to produce medical biodefense countermeasures. This task has not been given sufficient priority by DoD to produce the intended results. Furthermore, the disjointed and ineffective management and inadequate funding of current efforts are clear indications that DoD leaders lack an adequate grasp of the commitment, time, scientific expertise, organizational structure, and financial resources required for success in developing vaccines and other pharmaceutical products. Developing these products is a difficult endeavor, even with strong leadership and adequate resources. The fragmented half-measures of DoD's current effort cannot be expected to succeed.

RECOMMENDATIONS FOR ACTION

Maintaining the status quo in DoD only assures a long, costly, and perhaps fruitless wait for new vaccines and therapeutic products. The successful development and licensure of new countermeasures to protect against present and future biological warfare threats require a substantial and sustained effort, including having strong, scientifically knowledgeable leadership and adequate funding. To help ensure that DoD has an effective program to develop medical biodefense countermeasures to meet its unique needs, the committee makes recommendations in three areas:

1. Making the program a genuine priority
2. Establishing a sound infrastructure to support the program
3. Addressing other challenges related to the development of medical countermeasures

Key recommendations are discussed below, and the complete set of recommendations appears in Box ES-1 at the end of this chapter.

Making the Development of Medical Countermeasures a Priority

A decision by DoD and national leaders to make the DoD program to develop medical countermeasures against biological warfare agents a genuine priority is the essential first step to set the stage for an effective program aimed at meeting unique DoD needs.

To ensure that DoD has an effective research and development program for medical biodefense countermeasures, the committee makes the following recommendation:

1. The Secretary of Defense and Congress must make the DoD program for medical biodefense countermeasures a high priority.

Making the program a high priority will entail key changes:

- organizing the program to promote accountability and effective coordination throughout all phases of research, development, and product approval;
- installing leaders broadly knowledgeable in biotechnology with specific expertise in the development of vaccines and pharmaceutical products;
- supporting the development of a strong scientific infrastructure, including scientific personnel with expertise in pharmaceutical product development and facilities for research and animal testing; and
- providing the necessary funding to achieve program goals.

Organizing an Effective Program with Accountability for Performance

Assign Responsibility and Authority to a New DoD Agency

If the development of medical countermeasures is made a genuine priority, changes will be necessary to establish a sound infrastructure for integrated and comprehensive management of the research, development, and approval processes. After reviewing various organizational options, ranging from changes within the existing organizational framework to ending the DoD-based program (see Chapter 2), the committee concluded that the task requires the creation of a new DoD agency—designated by the committee as the Medical Biodefense[5] Agency—that consolidates the functions and resources of several existing activities (see Figure 2-2 for an organization chart).

2. Congress should authorize the creation of the Medical Biodefense Agency, a new DoD agency responsible for the research and development program for medical countermeasures against biological warfare agents. The committee recommends the following features for this agency:

- **It should report directly to the Under Secretary of Defense for Acquisition, Technology, and Logistics.**
- **The functions of existing medical biodefense programs should be transferred to the new Medical Biodefense Agency, along with their personnel and funding, including the medical biodefense component of the Chemical and Biological Defense Program (including units within the Army such as the U.S. Army Medical Research Institute of Infectious Diseases [USAMRIID]) and related activities in the Defense Advanced Research Projects Agency (DARPA). The research and development program for medical countermeasures against infectious diseases should also be transferred into the Medical Biodefense Agency.**
- **The Medical Biodefense Agency should function on the premise that to speed the development of countermeasures it is necessary to benefit fully from research and development efforts beyond DoD's intramural program so as to bring the expertise and creativity of industry and the academic community to the task.**
- **The Medical Biodefense Agency should ensure that DoD's medical biodefense activities are coordinated with and take full advantage of the related activities of NIH and that DoD's efforts are focused on meeting unique DoD needs.**

Essential functions of the new agency include the following:

- Establish and maintain knowledgeable leadership and effective management.

[5]Hereafter the term biodefense is used to describe defense against naturally occurring infectious disease as well as biowarfare agents.

- Offer effective identification, evaluation, and prediction of, as well as advocacy for, medical biodefense needs.
- Encourage and facilitate coordination with related efforts of other government agencies, the academic community, and the private sector.
- Seek necessary resources.
- Promote program stability.
- Understand and promote the use of the best science for the task.
- Tailor the acquisition process for medical countermeasures to use only FDA's regulatory requirements as the basis for assessing the technical merits of candidate products.
- Provide the means for obtaining expert advice on ethical and legal issues.

The agency should have a highly qualified director with strong experience in vaccine and drug research and development and manufacturing, including the rapidly evolving contributions of biotechnology. It is essential that the director have direct authority over the agency's budgeting and over its full range of management and operational activities, which should extend from basic research through full-scale production. An organizational approach that creates competing lines of authority and multiple reporting relationships, as the current matrix scheme does, is not adequate to address the multiple management and scientific challenges that DoD faces.

Of particular importance is ensuring that the Medical Biodefense Agency has the authority to manage the transition of candidate products from the science and technology stage into, and their progress through, the DoD acquisition system. In particular, the Medical Biodefense Agency should have the authority to use funds from science and technology accounts (e.g., budget activity 6.3) to support Phase 1 and even Phase 2 clinical trials before a candidate product is subject to acquisition system review.

The arbitrary separation between DoD's programs to develop medical countermeasures against biological warfare agents and against infectious diseases of military significance should be eliminated. These programs address similar scientific and technological questions and require closely related expertise and facilities. Also, with concerns about biological warfare threats expanding to include a wider range of naturally occurring and novel biological agents, the line between the two programs is becoming even less distinct and meaningful than it was in the past.

For its current scope, the DoD program to develop medical biodefense countermeasures is underfunded, based on the experience of other relevant government agencies and the private sector. However, the program should be better focused before any substantial increase in funding occurs. The Medical Biodefense Agency's budget should initially include as a baseline the funding currently allocated to the research and development activities for medical biological defense in the Chemical and Biological Defense Program ($189 million for FY 2003 in budget activities 6.1 through 6.5), funding for related activities in DARPA ($79 million for FY 2003), and funding for research and product development for the infectious disease program ($54 million for FY 2003). Related management support funding for each of these program areas (budget activity 6.6) should also be included.

In addition to this baseline of $322 million, the agency should receive an initial increase of $100 million, rising over the first 5 years to $300 million above the baseline amount. This increase reflects, in part, the expectation that more work will be done by civilian instead of military personnel and in non-DoD facilities, via a vibrant extramural program. The cost of salaries for military personnel and the operation of military facilities (e.g., USAMRIID) is presently covered in accounts of the military services, not the Chemical and Biological Defense Program. In addition, some candidate products are moving into later phases of development which traditionally are more costly. DoD and Congress should expect the new agency's funding needs to increase further as additional products reach this stage. This budget proposal does not include funds for the procurement of products after licensure.

Additional funding should be provided, as well, to renovate or replace the deteriorating and overcrowded USAMRIID facility to preserve the availability of its unique animal testing and holding space and laboratories equipped for research involving lethal pathogens and to ensure that it has the capacity to employ up-to-date technologies in research, testing, and evaluation.

The committee is strongly persuaded that creation of the Medical Biodefense Agency will be the most effective means of improving DoD's research and development program for medical biodefense countermeasures. As a result, much of the discussion and many of the recommendations throughout the report are framed in reference to this agency. The committee sees the strengths of its recommended approach as including preservation of DoD control over program priorities, integrated planning and management of all stages in the development of medical biodefense countermeasures, increased visibility of and priority for this work within DoD, increased expertise among the program leadership and managers, enhanced opportunity for coordination with related NIH work on bioterrorism countermeasures, and expanded access to contributions from extramural researchers. Disadvantages noted by the committee include the disruption of establishing a new agency and the potential difficulty of attracting a director and agency staff with the necessary qualifications (see Table 2-1).

In the event the Medical Biodefense Agency were not created, the need to establish a substantially more effective infrastructure for the DoD medical countermeasures program, as well as the need to address other critical challenges affecting that program, will remain and should be addressed by DoD. Those efforts can and should be guided by the same considerations that the committee discusses as the basis for its proposals for the Medical Biodefense Agency.

Establish External Oversight and Accountability for Performance

An independent, external review committee, composed of experts drawn from academia, the biotechnology and pharmaceutical industries, and other segments of the private sector who bring up-to-date scientific and managerial expertise in research and product development for vaccines and drugs, should be formed to monitor the performance of the DoD research and development program for medical biodefense countermeasures.

> **3. Congress should establish an external review committee of experts in the development of vaccines and drugs to review and evaluate the program and performance of the DoD research and development program for medical biodefense countermeasures each year. The committee should report its findings each year to the Secretary of Defense and the Congress.**

Maintaining DoD control over a program to develop medical biodefense countermeasures is particularly important to help ensure that unique DoD needs receive attention. However, DoD has failed to respond adequately to previous reports (e.g., IOM, 2002; Top et al., 2000) with similar recommendations for change. The committee believes that the development of medical biowarfare countermeasures requires the same urgency as the development of medical bioterrorism countermeasures; therefore if DoD does not take steps necessary to establish an effective program and make appropriate progress within 3 years, the committee recommends, as a last resort, transferring all or part of this responsibility from DoD to an agency responsible for promoting the development of medical countermeasures for bioterrorism defense. At present, NIH appears to be the best alternative because of its depth of scientific expertise and its substantial funding to support work on medical defenses against bioterrorism. However, NIH has little tradition of product development or history of focus on military-specific needs, and among many competing national public health priorities this additional task may not be given sufficiently high priority.

Other Challenges Requiring DoD Action

For the policy and organizational changes to have the positive impact the committee seeks, the Medical Biodefense Agency also has to find ways, often in collaboration with others, to overcome other substantial obstacles to the successful development and licensure of medical countermeasures. These include:

1. establishing effective collaborations with academia and industry;
2. meeting the challenges of the regulatory process, including helping to establish and maintain a strong scientific base for the evaluation of biodefense products;
3. enhancing the supply and effective use of resources needed for research and testing of biodefense products, including laboratory animals and animal facilities and specialized laboratories; and
4. ensuring the availability of a well-trained workforce.

Establishing Effective Collaborations with Academia and Industry

Partnerships with the academic community and with biotechnology and pharmaceutical companies will be crucial to the success of DoD's efforts to develop medical countermeasures. Early research and discovery leading to new candidates for vaccine and drug countermeasures should be a mixture of intramural and extramural work, depending on the leadership's assessment of the most effective way to achieve programmatic goals. With no federally owned facilities for full-scale manufacturing of vaccines or drugs, industry is an essential partner. Biotechnology and pharmaceutical firms also have expertise that can aid all phases of research and development. Deterrents to participation in these efforts include complex, cumbersome contracting procedures; the potential instability of government funding; and concerns about potential liability risks. Larger companies have been deterred by factors including short-term opportunity costs and little commercial market for many biodefense products.

To encourage increased involvement by academia and private sector firms in the development of medical countermeasures for DoD, the Medical Biodefense Agency should make full use of all available funding mechanisms, including "other transactions" authority, which is specifically intended for agreements with commercial firms that do not normally contract with DoD. In addition, DoD has new authority to follow simplified acquisition procedures and to make advance purchases of medical countermeasures that can be produced and delivered within 5 years, with the presumption, but no requirement, that those countermeasures can be licensed by FDA. DoD should make maximum use of its available authority to indemnify firms and others involved in developing these products. As soon as possible, Congress should extend the liability provisions in the Homeland Security Act for the smallpox vaccine to other medical biodefense countermeasures.

Meeting the Challenges of the Regulatory Process

As the regulatory authority reviewing data on the safety and efficacy of all medical products, FDA is another essential partner in the development of biodefense countermeasures. Interactions with FDA begin before the start of human testing of a candidate product and continue after a product is licensed and in use. The adoption of the Animal Efficacy Rule (FDA, 2002) removes a formidable barrier to licensure of medical biodefense countermeasures. Nonetheless, extensive research and testing will be needed to establish the scientific basis for applying this new regulatory mechanism. The Medical Biodefense Agency should cooperate with FDA, NIH, and others to make data on animal models readily available. In addition, FDA should collaborate with the scientific community in efforts to enrich the science base that it will have to draw on in order to apply the Animal Efficacy Rule.

To hasten action on medical countermeasures, FDA has adopted practices that are unusually proactive. However, these expanded efforts translate into the need for more staff or the diversion of staff from other tasks. Although FDA has already received some additional funding and personnel to support its additional work to respond to biowarfare and bioterrorism threats, Congress should ensure that funding continues to be sufficient to allow FDA to sustain these efforts.

It may be possible to speed DoD's access to certain medical countermeasures through use of existing FDA authority to approve products for use by a specific population (e.g., healthy adults) or under specific circumstances. Except under newly established provisions for emergency use, products still in investigational status (or not approved for a specific use) can be administered to military personnel only in accordance with informed consent procedures or with a presidential waiver of those procedures. Ensuring that DoD can respond in an effective and timely manner to any need for emergency use of medical countermeasures will require ongoing planning and coordination among various components within the department.

Overcoming Current and Potential Bottlenecks Related to Research Resources

Research to identify candidate countermeasures against biological warfare agents and the subsequent work necessary to bring products to licensure will require extensive use of animal models (and, thus, the facilities to house animals) and specialized laboratory facilities with appropriate biosafety features. Nonhuman primates, especially Indian-origin rhesus macaques, are in high demand for this and other types of research. In addition, clinical testing requires access to facilities that can produce small supplies of candidate countermeasures in compliance with FDA's current Good Manufacturing Practice (cGMP).

The Medical Biodefense Agency should participate in a broad-based assessment of the likely demand for nonhuman primates and other laboratory animals, animal facilities (for testing and housing), and GMP production of candidate products necessary to support current or planned research on all types of biodefense countermeasures. The agency should promote and participate in efforts to encourage the use of nonhuman primates other than Indian-origin rhesus macaques and to assess and coordinate the use of nonhuman primates and the use of government-owned testing facilities for biodefense research and product development. In addition, the committee recommends that DoD provide funding to carry out renovations necessary to ensure that USAMRIID has fully functional biosafety level 3 and 4 (BSL-3 and BSL-4) facilities for laboratory and animal research.

Ensuring the Availability of a Well-Trained Workforce

The nation faces a limited supply of scientific and technical personnel with the expertise needed for work on medical countermeasures (Partnership for Public Service, 2003). The Medical Biodefense Agency should define needed workforce capabilities and aid in the development and implementation of training programs designed to meet those needs. In addition, to attract and retain a skilled workforce, the agency should use DoD's newly available authority to offer salaries that are more competitive with those in academia and industry.

BOX ES-1
Recommendations

To ensure that DoD has an effective research and development program for medical biodefense countermeasures, the committee makes the following recommendations:

Making Medical Countermeasures a Priority

1. The Secretary of Defense and Congress must make the DoD program for medical biodefense countermeasures a high priority.

Organizing an Effective Program with Accountability for Performance

2. Congress should authorize the creation of the Medical Biodefense Agency, a new DoD agency responsible for the research and development program for medical countermeasures against biological warfare agents. The committee recommends the following features for this agency:

- It should report directly to the Under Secretary of Defense for Acquisition, Technology, and Logistics.
- The functions of existing medical biodefense programs should be transferred to the new Medical Biodefense Agency, along with their personnel and funding, including the medical biodefense component of the Chemical and Biological Defense Program (including units within the Army such as USAMRIID), and related activities in the Defense Advanced Research Projects Agency. The research and development program for medical countermeasures against infectious diseases should also be transferred into the Medical Biodefense Agency.
- The Medical Biodefense Agency should function on the premise that to speed the development of countermeasures it is necessary to benefit fully from research and development efforts beyond DoD's intramural program so as to bring the expertise and creativity of industry and the academic community to the task.
- The Medical Biodefense Agency should ensure that DoD's medical biodefense activities are coordinated with and take full advantage of the related activities of NIH and that DoD's efforts are focused on meeting unique DoD needs.

3. Congress should establish an external review committee of experts in the development of vaccines and drugs to review and evaluate the program and performance of the DoD research and development program for medical biodefense countermeasures each year. The committee should report its findings each year to the Secretary of Defense and the Congress.

4. If the review committee finds that after a 3-year period of operation the DoD research and development program for medical biodefense countermeasures has failed to make progress that the committee considers

continued

BOX ES-1 Continued

appropriate, Congress should transfer from DoD, in part or in whole, responsibility for the development of medical biodefense countermeasures and reassign that responsibility to an agency responsible for promoting the development of medical countermeasures for bioterrorism defense, such as the NIH or another agency considered appropriate.

Establishing Effective Collaboration with Academia and the Private Sector

5. The Medical Biodefense Agency should fully utilize "other transactions" authority as a means of encouraging academia and private sector firms to participate in the research and development of medical biodefense countermeasures to meet DoD needs.

6. Congress should authorize the Medical Biodefense Agency to sign multiyear contracts without a requirement for full, up-front funding of any termination liabilities.

7. DoD and DHHS should make maximum permissible use of statutory indemnification authority under existing legislation to encourage entities in the private sector, including universities and other research institutions and companies, to enter into agreements to develop and manufacture medical countermeasures against biowarfare agents. As soon as possible, legislation should be enacted creating a system comparable to that for the smallpox vaccine under the Homeland Security Act, under which suits for personal injuries allegedly caused by biowarfare countermeasures may be brought only against the federal government, which would retain the right to recover damages resulting from such suits from manufacturers or other covered persons if their misconduct (gross negligence, illegal acts, willful misconduct, or violation of government contract obligations) was shown to be the cause of the injuries.

Meeting the Challenges of the Regulatory Process

8. The Medical Biodefense Agency and NIH should cooperate in making information on animal models relevant for the development of medical biowarfare countermeasures available to qualified investigators. The DoD agency should work with NIH and engage FDA to develop additional animal models that will be useful for specific agents or products of particular concern to DoD. The Medical Biodefense Agency should receive funding specifically for this task.

9. FDA should work with the scientific community to enrich the science base that the agency will have to draw on in order to apply the Animal Efficacy Rule. FDA should receive sufficient funding to support both intramural and extramural work on these issues.

10. Congress should ensure that adequate funding is provided to support the additional work that FDA is carrying out in response to threats from bioterrorism and biowarfare.

Overcoming Current and Potential Resource Bottlenecks

11. The Medical Biodefense Agency should participate in a national effort to support the maintenance and expansion of nonhuman primate research resources, which will be critical to the success of efforts to develop medical biodefense countermeasures. The Medical Biodefense Agency should be provided with sufficient funding for these activities.

12. The Medical Biodefense Agency should participate in interdepartmental efforts to make a formal assessment of the need for facilities for animal testing and holding and for GMP-compliant manufacturing of material for clinical testing that will arise from research efforts to develop medical countermeasures to biowarfare or bioterrorism agents that are under way, planned, or likely.

13. The Medical Biodefense Agency should promote the development of, and participate in a system for prioritizing the use of, specialized government-owned testing facilities that are essential for research and development of medical biodefense countermeasures.

14. DoD should provide funding to carry out the renovations necessary to ensure that USAMRIID can continue operation of fully functional BSL-3 and BSL-4 facilities for laboratory and animal research.

Ensuring the Availability of a Well-Trained Workforce

15. The Medical Biodefense Agency should define the capabilities needed for its medical countermeasures workforce, collaborate with NIAID and industry to develop a training curriculum, and support training programs in areas of special expertise needed for research and development of medical countermeasures. The Medical Biodefense Agency could contribute unique DoD resources in areas of aerobiology and the development of animal models of human diseases caused by biological warfare agents.

16. DoD should use its authority under the National Defense Authorization Act for FY 2004 (P.L. 108-136) to offer more competitive salaries to technical experts to bring necessary expertise in biotechnology and pharmaceutical research and development to the Medical Biodefense Agency. Budgeting for the Medical Biodefense Agency should reflect the need to use such provisions to recruit experienced scientific and technical personnel.

REFERENCES

Bush GW. 2002. Remarks by the President at the signing of the Public Health Security and Bioterrorism Preparedness and Response Act of 2002, June.

Cohen WS. 1997. Message of the Secretary of Defense. In *Proliferation: Threat and Response*. Washington, DC: Department of Defense. [Online]. Available: http://www.defenselink.mil/pubs/prolif97/message.html [accessed September 24, 2003].

Defense Science Board. 2001. 2000 Summer Study Executive Summary, Volume 1. In *Protecting the Homeland*. Washington, DC: Office of the Under Secretary of Defense for Acquisition, Technology, and Logistics.

Defense Science Board. 2002. *2001 Summer Study on Defense Science and Technology*. Washington, DC: Office of the Under Secretary of Defense for Acquisition, Technology, and Logistics.

DHHS (Department of Health and Human Services). No date. *FY 2004 Budget in Brief*. Washington, DC: Department of Health and Human Services. [Online]. Available: http://www.dhhs.gov/budget/04budget/fy2004bib.pdf [accessed March 5, 2003].

DiMasi JA, Hansen RW, Grabowski HG. 2003. The price of innovation: new estimates of drug development costs. *Journal of Health Economics* 22(2):151–185.

DoD (Department of Defense). 1993. *Department of Defense Directive: DoD Immunization Program for Biological Warfare Defense*. Number 6205.3. Washington, DC: Department of Defense.

DoD. 2000. *Department of Defense Directive: Use of Investigational New Drugs for Force Health Protection*. Number 6200.2. Washington, DC: Department of Defense.

DoD. 2003. *Department of Defense Chemical and Biological Defense Program. Volume I: Annual Report to Congress*. Washington, DC: Department of Defense. [Online]. Available: http://www.acq.osd.mil/cp/nbc03/vol1-2003cbdpannualreport.pdf [accessed June 27, 2003].

Evans D. 2003. Updated budget charts. E-mail to L. Joellenbeck, Institute of Medicine, Washington, DC, June 24.

Fauci AS. 2003. NIAID biodefense research and collaborations with DoD. Presentation to the Institute of Medicine and National Research Council Committee on Accelerating the Research, Development, and Acquisition of Medical Countermeasures Against Biological Warfare Agents, Meeting III. Washington, DC.

FDA (Food and Drug Administration). 1999. *From Test Tube to Patient: Improving Health Through Human Drugs*. Rockville, MD: Food and Drug Administration, Center for Drug Evaluation and Research. [Online]. Available: http://www.fda.gov/cder/about/whatwedo/testtube-full.pdf [accessed July 25, 2003].

FDA. 2002. New drug and biological drug products: Evidence needed to demonstrate effectiveness of new drugs when human efficacy studies are not ethical or feasible. Final rule. 21 C.F.R. Parts 314 and 601. *Federal Register* 67(105):37988–37998.

Grant C. 2003. Aventis Pasteur Vaccine Development. Handout to Institute of Medicine and National Research Council Committee on Accelerating the Research, Development, and Acquisition of Medical Countermeasures Against Biological Warfare Agents, Meeting III. Washington, DC.

IOM (Institute of Medicine). 2002. *Protecting Our Forces: Improving Vaccine Acquisition and Availability in the U.S. Military*. Lemon SM, Thaul S, Fisseha S, O'Maonaigh HC, eds. Washington, DC: The National Academies Press.

IOM. 2003. *Financing Vaccines in the 21st Century: Assuring Access and Availability*. Committee on the Evaluation of Vaccine Purchase Financing in the United States. Washington, DC: The National Academies Press.

Linden C. 2002. Sponsor presentation on the study charge. Presentation to the Institute of Medicine and National Research Council Committee on Accelerating the Research, Development, and Acquisition of Medical Countermeasures Against Biological Warfare Agents, Meeting I. Washington, DC.

NVAC (National Vaccine Advisory Committee). 1999. Lessons learned from a review of the development of selected vaccines. *Pediatrics* 104(4):942–950.

Partnership for Public Service. 2003. *Homeland Insecurity: Building the Expertise to Defend America from Bioterrorism.* Washington, DC: Partnership for Public Service. [Online]. Available: http://www.ourpublicservice.org/publications3735/publications_show.htm?doc_id=181630 [accessed July 8, 2003].

Perry WJ. 1996. Preface: the new threat from nuclear, biological, and chemical weapons. In *Proliferation: Threat and Response.* Washington, DC: Department of Defense. [Online]. Available: http://handle.dtic.mil/100.2/ADA314341 [accessed September 24, 2003].

PhRMA (Pharmaceutical Research and Manufacturers of America). 2000. Why Do Prescription Drugs Cost So Much and Other Questions About Your Medicines. [Online]. Available: http://www.phrma.org/publications/publications/questions/questions.pdf [accessed July 2, 2003].

Public Citizen. 2001. *Rx R&D Myths: The Case Against the Drug Industry's R&D "Scare Card."* Washington, DC: Public Citizen's Congress Watch. [Online]. Available: http://www.citizen.org/publications/release.cfm?ID=7065&secID=1078&catID=126 [accessed June 23, 2003].

Struck MM. 1996. Vaccine R&D success rates and development times. *Nature Biotechnology* 14(5):591–593.

Top FH Jr., Dingerdissen JJ, Habig WH, Quinnan GV Jr., Wells RL. 2000. DoD Acquisition of Vaccine Production: Report to the Deputy Secretary of Defense by the Independent Panel of Experts. In DoD. 2001. *Report on Biological Warfare Defense Vaccine Research and Development Programs.* Washington, DC: Department of Defense. [Online]. Available: http://www.acq.osd.mil/cp/bwdvrdp-july01.pdf [accessed February 19, 2004].

U.S. Congress, House Armed Services Committee. 1993. *Countering the Chemical and Biological Weapons Threat in the Post-Soviet World.* Report of the Special Inquiry into the Chemical and Biological Threat. Committee print 15. 102nd Congress, 2nd Session. February 23.

U.S. Congress, Office of Technology Assessment (OTA). 1993. *Pharmaceutical R&D: Costs, Risks and Rewards.* OTA-H-522. Washington, DC: U.S. Government Printing Office. [Online]. Available: http://www.wws.princeton.edu/~ota/ns20/year_f.html [accessed July 8, 2003].

White House. 2003. President details Project BioShield. Washington, DC, February 3. [Online]. Available: http://www.whitehouse.gov/news/releases/2003/02/print/20030203.html [accessed February 5, 2003].

1

Ending Half-Measures for Countermeasures: The Challenge and Major Recommendations

The biodefense efforts of the Department of Defense (DoD) are poorly organized to develop and license vaccines, therapeutic drugs, and antitoxins to protect members of the armed forces against biological warfare agents.

The development and licensure of new vaccines and drugs is a difficult, expensive, and time-consuming process. Moreover, biodefense products pose special scientific, regulatory, and ethical challenges because it is generally unacceptable to expose humans to biowarfare agents to establish the efficacy of those products. Accelerating the development and licensure of such products will require strong and creative scientific leadership and a sustained commitment of adequate financial and other resources. Current DoD efforts, however, are characterized by fragmentation of responsibility and authority, changing strategies that have resulted in lost time and expertise, and a lack of financial commitment commensurate with the requirements of program goals. These factors, together with high regulatory hurdles for obtaining Food and Drug Administration (FDA) approval,[1] mean that since the Gulf War of 1990–1991 DoD has gained no new vaccines and only a few drugs to protect its military personnel against biological warfare agents.

---

[1]It was not possible to license new vaccines or drugs against biological warfare agents until July 2002, when the FDA's "Animal Efficacy Rule" became effective (FDA, 2002). The Animal Efficacy Rule allows the use of efficacy data from animal studies when tests of efficacy in humans are not ethical or feasible, as is generally the case with medical biodefense countermeasures.

This serious situation exists despite declarations by Presidents, Secretaries of Defense, congressional committees, and advisory groups (Bush, 2002; Clinton, 1994; Cohen, 1997; Defense Science Board, 2001, 2002; Perry, 1996; U.S. Congress, House Armed Services Committee, 1993) that biological warfare poses a significant threat to the safety and effectiveness of the nation's armed forces, despite targeted DoD programs initiated in 1998 and 2002 to vaccinate large numbers of military personnel against anthrax and smallpox (Chu, 2002; Cohen, 1998), and despite a DoD commitment to acquire vaccines against all validated biological warfare threats (DoD, 1993). Moreover, there is concern that advances in bioengineering will make possible the rapid introduction of new biological threats that may prove even more challenging to counter than the already serious threats posed by naturally occurring organisms (Defense Science Board, 2002; MacKenzie, 2003).

This report presents the findings and recommendations of a committee of the Institute of Medicine (IOM) and the National Research Council (NRC) convened to conduct a study mandated by Congress.[2] The charge to the committee is to recommend strategies for accelerating the DoD research, development, and licensure processes for new medical biodefense countermeasures.[3] Based on the study charge to address strategies for accelerating these processes, the committee focused its attention on the organization and management of these processes, rather than on details of specific scientific approaches. The committee was not asked to assess the nature or extent of any biological warfare threat or to compare the value to DoD of developing medical countermeasures against biological warfare agents relative to the pursuit of its other obligations. The committee viewed its task as resting on the premise that biological weapons pose a genuine threat to the health of military personnel, and therefore additional FDA-licensed medical countermeasures are urgently needed.

THE CURRENT CONTEXT FOR THE DEVELOPMENT OF MEDICAL COUNTERMEASURES

The context in which DoD is working to develop new vaccines and drugs to counter biowarfare agents has changed during the past 10 years, especially since 2001 (see Box 1-1). Continuing scientific and technological

---

[2]The study was called for in the National Defense Authorization Act for Fiscal Year 2002, P.L. 107-107 (2001). The study charge, the expertise of the committee, and the committee's approach to the study are discussed in the Preface.

[3]In this report, the term "licensed" is used to connote approval by FDA of either a Biologics License Application or New Drug Application.

advances in molecular biology, genomics, combinatorial chemistry, and understanding of microbial structure and replication are expanding the range of potential biological threats as well as offering opportunities to identify new countermeasures. For example, antibiotic resistance can be readily engineered into microbes, and some alterations could hinder the effectiveness of some vaccines (MacKenzie, 2003). Even so, the potential for new bioengineered pathogens does not eliminate the need for countermeasures against more familiar biological threats and continued work on products currently in development pipelines. In addition, advances in biotechnology and scientific understanding are facilitating the exploration of new types of countermeasures such as broad-spectrum antibiotics and antivirals, as well as possibilities for multivalent vaccines and rapid development of antibodies.

For decades, DoD carried out the nation's only significant effort to develop medical biodefense countermeasures. Now, however, a substantial biodefense research effort is under way within the Department of Health and Human Services. The government has responded to the experience of domestic bioterrorism with a major increase in funding for the National Institute of Allergy and Infectious Diseases (NIAID) of the National Institutes of Health (NIH)—initially $1.7 billion for fiscal year (FY) 2003 and $1.6 billion requested for FY 2004—to support new research and the renovation or construction of special laboratory facilities necessary for research involving biological threat agents (DHHS, no date). In addition, the proposal by President Bush for "Project BioShield" aims to create incentives for the pharmaceutical industry to manufacture and license medical countermeasures by making up to $6 billion available over the next 10 years to purchase those products for a national stockpile (White House, 2003). In contrast, DoD's funding for its research and development program for medical countermeasures against at least 10 biological agents amounts to only $267 million for FY 2003 (DoD, 2003a; Evans, 2003).[4]

The upsurge in funding and effort aimed at protecting the civilian population against bioterrorism will undoubtedly result in the development of new technologies and products that can also aid in protecting military personnel against the risks of biological warfare. Although there is considerable overlap in potential threat agents for the battlefield and

---

[4]DoD funding for FY 2003 includes program elements for medical biological defense in the Chemical and Biological Defense Program (under budget activities 6.1–6.5) and biomedical components of the biological warfare defense program of the Defense Advanced Research Projects Agency. The funding totals for medical biodefense exclude costs presently covered in accounts of the military services for salaries and benefits for military personnel and for operating certain facilities.

BOX 1-1
Events Related to the Development of Medical Biodefense Countermeasures

November 1969	United States renounces use of biological weapons
April 1972	Medical protection functions of Biological Defense Research Laboratory at Ft. Detrick transferred from the Army Materiel Command to USAMRIID, under the Army Medical Department
March 1975	United States ratifies Biological and Toxin Weapons Convention
May 1985	DoD Directive 5160.5 reaffirms the Department of the Army as the executive agent for research and development for chemical and biological defense
August 1990–July 1991	Gulf War deployment; licensed anthrax vaccine administered to U.S. troops; Botulinum toxoid vaccine in Investigational New Drug (IND) status administered without formal process of informed consent
November 1993	P.L. 103-160 (50 U.S.C. 1522) mandates overall coordination and integration of the chemical and biological warfare defense program (both medical and nonmedical components) by a single office within the Office of the Secretary of Defense; oversight is to be exercised through the Defense Acquisition Board process
November 1993	DoD Directive 6205.3 calls for developing the capability to acquire vaccines against all validated biological warfare threats
1994	Joint Service Agreement for Joint Nuclear, Biological, and Chemical Defense Management; Joint Program Office for Biological Defense chartered (April)

August 1995	Iraq reports weaponization of biological agents to United Nations Special Commission
October 1996	FDA Vaccines and Related Biological Products Advisory Committee provides guidance to DoD on the types of human antibody and animal challenge studies to conduct to support the effectiveness of a botulinum toxoid vaccine
December 1996	Joint Vaccine Acquisition Program Management Office established
November 1997	Prime contract awarded by DoD to DynPort Vaccine Company for development and licensure of biodefense vaccines
September 1999	10 U.S.C. 1107 and Executive Order 13139: Administration to members of the armed forces of INDs or drugs used for purposes not approved by the FDA must adhere to requirements for informed consent unless a waiver is granted by the President
September–October 2001	Distribution of anthrax spores through the U.S. postal system; five deaths result
July 2002	FDA "Animal Efficacy Rule" goes into effect (21 C.F.R. Parts 314 and 601: New Drug and Biological Drug Products; Evidence Needed to Demonstrate Effectiveness of New Drugs When Human Efficacy Studies Are Not Ethical or Feasible)
February 2003	Charter issued for Joint Requirements Office for Chemical, Biological, Radiological, and Nuclear Defense (JRO-CBRN)
April 2003	Implementation plan issued for management of the Chemical Biological Defense Program (transfers management of all science and technology activities to the Defense Threat Reduction Agency)

the civilian community, the perceived risks posed by these agents can be different in the two settings. Furthermore, there are important differences between the planning for protection against bioterrorism versus biological warfare (DoD, 1993, 2000; Fauci, 2003; Linden, 2002). The military has considered vaccination to be the primary medical strategy for battlefield protection of a defined and relatively small population. Mass vaccination of the civilian population against a range of potential biological threats is less appropriate and much less feasible. For DoD, the aim is to protect service members in a manner that allows them to maintain their combat effectiveness and limits the need for medical personnel and equipment to treat casualties. In addition, DoD seeks to deter and defeat the use of biological weapons by having troops medically protected. Having the capacity for rapid diagnosis and postexposure treatment is also essential for DoD, but it is less desirable as a primary strategy for protecting troops on the battlefield than it is for responding to a bioterrorism event in the civilian population.

The military and civilian biodefense programs also differ in their approach to achieving the development, FDA licensure, and manufacture of medical countermeasures. The aim of efforts supported by the increased funding for NIAID is to help ensure that new candidate countermeasures are discovered and developed to a stage at which initial studies of safety and efficacy are promising (i.e., Phase 1 or 2 clinical trials). The provisions for government purchases for a national stockpile as proposed under Project BioShield are intended to provide sufficient financial incentive for commercial firms to undertake the work necessary to complete the testing and licensure process for products that they manufacture. The DoD approach is to support the initial research to identify candidate countermeasures and to include in the terms of its cost-plus-award-fee prime systems contract an explicit requirement for the development and delivery of FDA-licensed products that it can administer to military personnel.

What has not changed are the challenges in developing any new vaccine or drug, including the cost and the substantial risk of failure, even in late stages of the process. The congressional Office of Technology Assessment (OTA) in 1993 estimated the average cost of bringing a new drug to licensure to be $237 million (1990 dollars) (U.S. Congress, OTA, 1993). More recent estimates have ranged from $110 million to $802 million (2000 dollars) (DiMasi et al., 2003; Public Citizen, 2001). As few as one candidate in 5,000 reaches clinical testing, and only 20 percent of candidates that begin clinical testing reach licensure (FDA, 1999; NVAC, 1999; PhRMA, 2000; Struck, 1996). Such estimates are based primarily on data for new drugs; equivalent estimates for vaccines and other biologics are rarely presented (IOM, 2003).

The process is also time consuming (see Box 1-2 for a brief outline of the process). An industry estimate of from 7 to 12 years for vaccine discovery and development was presented to the committee (Grant, 2003). For five vaccines reviewed by the National Vaccine Advisory Committee, the time from the beginning of Phase 1 trials to licensure ranged from 2 to 21 years (NVAC, 1999). FDA (1995) has cited a range of 4 to 20 years for drug development.

New techniques made possible by scientific advances in fields such as genomics, proteomics, and high-throughput screening are likely to speed the discovery of some candidate countermeasures, but are unlikely to accelerate some of the most time-consuming parts of the product development process, including the crucial assessment of product safety in human volunteers and efficacy based on animal data under new FDA regulatory guidelines (the "Animal Efficacy Rule") (FDA, 2002). FDA licensure of vaccines and drugs requires the submission of data demonstrating the efficacy of the product under review. For biodefense countermeasures, however, efficacy studies in humans are generally not feasible or ethical, presenting a barrier to licensure. FDA first provided guidance on the use of efficacy data from animal studies in combination with data on immune responses in humans in 1996 (Wykoff, 1998) and finalized regulations on the use of animal data in 2002 (FDA, 2002). The new regulations have now opened the path to licensure for new biodefense countermeasures; however, the use of animal-based efficacy testing is still an unfamiliar process and thus likely to require considerable time and effort to become regularized.

PROBLEMS HINDERING THE DOD EFFORT

Although DoD has maintained a research base for medical biodefense countermeasures for many years when few others were working in this field, the committee views the ineffective and inadequate organization and funding of the medical biodefense component of the Chemical and Biological Defense Program as a clear indication that DoD leaders lack an adequate grasp of the commitment, time, scientific expertise, and financial resources required for success in developing vaccines and other pharmaceutical products. Repeated changes in organization and strategy have not addressed those deficiencies, generating instead a flux that has ultimately resulted in disjointed and ineffective management.

Fragmentation of Responsibility and Authority

Successful development and licensure of pharmaceutical products, especially vaccines, requires knowledgeable oversight and the creation of an integrated and accountable development team whose collective exper-

BOX 1-2
Factors in the Pace of Countermeasure Development and Licensure

The pace of efforts to develop and license vaccines, therapeutic drugs, and antitoxins to protect against biological warfare agents is determined by progress during four broad stages: discovery, early development, advanced development (clinical testing and scale-up of manufacturing), and licensure of the product by the FDA. Following licensure, these products are also likely to be subject to requirements for postmarketing surveillance because of the need to validate their efficacy in humans and to accumulate additional evidence on their safety. The discussion below uses an industry estimate for the time required at each stage in the development of a vaccine (Grant, 2003). Any given product might progress more slowly or quickly than these estimates.

Discovery: 2 to 3 years Identification of a candidate drug or vaccine antigen draws on basic research regarding a pathogen's mechanism of action and the host's response to infection. An animal model for the disease is often a valuable research tool. Additional research identifies potential molecular targets for drug or antitoxin development or antigens that might be appropriate for a vaccine. Advances over the past decade in molecular biology, genomics, and proteomics are helping accelerate discovery with techniques for rapid screening of large numbers of antigens or chemical compounds such that a substance may be identified for further evaluation in less than a year.

Early Development: 2 to 3 years A potential new drug or vaccine antigen is produced in limited quantities for laboratory and animal testing to establish proof of concept—that the drug or vaccine candidate can be administered safely and shown to block the action of a pathogen or generate a protective immune response. A candidate product must be characterized and initial manufacturing processes must be developed and documented. Many candidates are abandoned as information accrues regarding their behavior in living systems and the feasibility of manufacturing them. Given sufficient priority and with adequate personnel and resources, the pace at this stage is set by scientific approaches and technical hurdles.

For candidate products to be taken on to human testing, an acceptable IND application must be filed with FDA. The application must include data on safety and biological activity from laboratory and animal tests; manufacturing processes; standards for establishing the safety, purity, potency, and consistency of pilot lots for human use; and detailed plans for the initial clinical testing. FDA must be apprised of the initiation of each stage of human testing. Data from those tests must be submitted to FDA as product development and testing proceeds. Human testing may go forward as long as FDA raises no objections.

Clinical testing begins with Phase 1 studies, which are usually carried out with fewer than 100 people and provide initial data on the safety, pharmacokinetics (how the body handles a product), and if measurable, biological activity of a candidate product. Factors affecting the rate at which clinical studies proceed include the quality and organization of the documents submitted to FDA, the pace of approvals by institutional review boards, the rate of recruitment of study participants, and the period of time over which the action of the product must be evaluated.

Advanced Development: 2 to 3 years Products that remain in consideration proceed to Phase 2 studies, for tests in a few hundred people. Phase 2 testing for traditional drug candidates provides data on safety and biological activity at different dosage levels in patients. Phase 3 studies usually involve no fewer than 5,000 participants and provide definitive data on safety and efficacy. For most biodefense products, ethical considerations make it impossible to conduct human studies to demonstrate efficacy. Instead, evidence of efficacy in animals must be correlated with surrogate markers of the human response to the countermeasure. The time that will be required for this newly established pathway to product approval is uncertain. FDA requirements for specific types of testing and for review of data submissions and manufacturing processes also affect development time.

Manufacturing processes must be scaled up to full production levels and shown to be reproducible. For vaccines, several years of effort may be required to establish final manufacturing procedures because the biological processes involved are subject to inherent variability. Chemical processes to produce drugs usually have less variability and thus can be developed more rapidly.

Preparation and Review of Product License Application: 1 to 3 years At the conclusion of clinical testing and scale-up of production, an application is submitted to FDA for product licensure. FDA determines whether the data are sufficient to demonstrate the safety and efficacy of the candidate product and to show that the manufacturing and testing procedures are satisfactory. License applications for products designated as priority or fast track are normally acted on within 6 months. The pace of FDA action is affected by such factors as the quality and completeness of the data submissions, the overall numbers of products under review, and the availability of staff to perform reviews.

tise includes basic science, FDA regulatory matters, animal and human testing, development of manufacturing processes, production of pilot lots, and full-scale manufacturing. Despite nominally centralized oversight of the Chemical and Biological Defense Program within the Office of the Secretary of Defense, the DoD effort is in practice fragmented among multiple chains of command and burdened by organizational complexity. Research is now overseen by the Defense Threat Reduction Agency (DTRA) and executed predominantly by personnel under the U.S. Army Medical Command, while advanced development is managed by the Joint Program Executive Office for Chemical and Biological Defense under the Army Acquisition Executive. Program requirements are established within the Joint Chiefs of Staff, and budgets are dispersed among all of these organizations and their chains of commands (who often perceive other uses of these resources as being of higher priority). This arbitrary divide between research and advanced development—despite being the standard model for the development of weapons systems—creates a diffusion of authority and responsibility that has resulted in inefficiencies, inadequate funding, and a lack of accountability for efforts to develop vaccines and drugs for medical biodefense.

Losing Time and Expertise Through Organizational Adjustments and Changing Strategies

Over the past decade, the Chemical and Biological Defense Program has been subject to repeated reorganizations intended to make it function more effectively. These reorganizations have entailed creating brand new organizational units or giving existing units new and unfamiliar responsibilities for managing research and development for vaccines and other pharmaceutical products.

In 1993, Congress required that all of DoD's chemical and biological defense activities, both medical and nonmedical, be overseen by a single office within the Office of the Secretary of Defense.[5] This responsibility was assigned to what was then the Assistant to the Secretary of Defense for Atomic Energy (since redesignated as the Assistant to the Secretary of Defense for Nuclear and Chemical and Biological Defense Programs [ASTD(NCB)]). At the same time, Congress also directed that oversight of the program be exercised through the Defense Acquisition Board process. This requirement resulted in the transfer of responsibility for advanced

[5]National Defense Authorization Act for Fiscal Year 1994, P.L. 103-160 (1993, 50 U.S.C. 1522).

development of medical countermeasures from the U.S. Army Medical Command, which had successfully licensed vaccines developed through the Army's infectious disease research program, to the newly created Joint Program Office for Biological Defense, which had responsibility for both medical and nonmedical products. In 1996, the Joint Vaccine Acquisition Program was established within the Joint Program Office specifically to manage the advanced development and licensure of candidate vaccines by a prime systems contractor.

During the spring of 2003, a reorganization affected all phases of the Chemical and Biological Defense Program. In particular, a new Joint Requirements Office for Chemical, Biological, Radiological, and Nuclear Defense was established. Also, responsibility for management of research and early product development was transferred from the U.S. Army Medical Command to DTRA, an organization that has historically focused primarily on nuclear threats. DTRA has little resident expertise in managing or conducting the biomedical research necessary for the development of vaccines, drugs, and other medical countermeasures against biowarfare agents.

In addition, since the early 1990s the DoD strategy for producing biodefense vaccines has changed three times in response to concerns about affordability and cost-effectiveness (Johnson-Winegar, 2000). Initial planning for a government-owned, contractor-operated vaccine production facility gave way to investigation of a contractor-owned, contractor-operated approach. The strategy adopted in the mid-1990s, and currently being followed, is the use of a prime systems contract, which was awarded in late 1997. From the time responsibility for advanced development was removed from the U.S. Army Medical Command until the prime systems contract was awarded, DoD had no mechanism in place through which to pursue the advanced development of candidate countermeasures, and even with the prime systems contract in place, the first clinical trials did not begin until 2000.

With these repeated reorganizations and shifts in approach and with no one official clearly responsible for overall program results, time has been lost again and again while new units are organized and begin to recruit staff with relevant expertise, or worse yet, operate without adequate expertise. The willingness of upper-level decision makers within DoD to repeatedly make such changes indicates to the committee an awareness of the problem but a lack of understanding of the level of experience, expertise, and leadership, as well as the organizational imperatives, necessary to shepherd candidate vaccines and drugs through the long and difficult research, development, and licensure process.

Lack of Financial Commitment Commensurate with the Requirements of Program Goals

Insufficient and unstable funding offers further evidence that the research and development program for medical countermeasures has not been given sufficient priority by DoD. The medical component of DoD's biodefense program covers work on numerous vaccines, drugs, antitoxins, and diagnostics to protect against more than 10 potential biological threats (see Table A-2). As recently as 1997, DoD funding for medical biodefense was less than $50 million (see Table 1-1).[6] Funding rose, reaching $304 million in FY 2002, for the medical biodefense portion of the budgets for the Chemical and Biological Defense Program and the Defense Advanced Research Projects Agency (DARPA), but it has since declined.

Short- and long-term planning decisions have reallocated funds in ways that have disrupted or delayed work on the development of individual products. For example, funding for DoD's work on a tularemia vaccine was interrupted by programming decisions for FY 2004 and 2005 that would have effectively halted work on the vaccine without assistance from NIAID. Long-term prospects for that vaccine are now uncertain. In addition, the committee is concerned about the lack of a clear plan to provide the financial resources needed to replace or renovate the deteriorating, outmoded, and overcrowded facility, now more than 30 years old, housing the laboratories of the U.S. Army Medical Research Institute of Infectious Diseases (USAMRIID). USAMRIID is the principal site for DoD's medical biodefense research and a unique resource for animal testing under high-level biosafety conditions, which is necessary at many stages throughout the development process for medical countermeasures.

When the Congress, the President, or DoD identifies a program as a high priority, tremendous resources can be brought to bear, as with the $1.7 billion directed to the NIAID research program for FY 2003 and plans for similar amounts in coming years. Within DoD, annual funding for the Missile Defense Agency, which has a different but also difficult research and development task with a high risk of failure, has been built up from almost $3 billion to more than $7.6 billion over the past several years (see Figure 1-1).

[6]The funding totals for medical biodefense exclude costs presently covered in accounts of the military services for salaries for military personnel and for operation of facilities such as USAMRIID.

Impediments Beyond DoD

Despite emphasizing the adverse effect of DoD's failure to make the development and licensure of medical countermeasures a sufficiently high priority, the committee recognizes that the lack of progress in efforts to develop new biodefense vaccines and drugs was due, in part, to other serious impediments that lie outside DoD's control. One of the most important problems was the impossibility of establishing the efficacy of these products in humans, given the lethal nature of biological warfare agents and the inability to conduct studies on the basis of natural exposures of human populations to these pathogens. As the committee has noted, the adoption in 2002 of the Animal Efficacy Rule (FDA, 2002) removes a formidable barrier to FDA licensure of medical biodefense countermeasures. However, the application of this new regulatory mechanism will require extensive (and currently unfunded) research and testing.

Further, facilities for the production of test lots of candidate countermeasures (especially vaccines) in compliance with Good Manufacturing Practice (GMP) standards are limited, and the regulations regarding the use of these facilities are increasingly complex. The nation also faces a limited supply of scientific and technical personnel with the expertise required to carry out the work (Partnership for Public Service, 2003) and a limited supply of nonhuman primates and specialized laboratory facilities necessary for testing candidate products (NRC, 2003; Parker, 2003). In addition, a new emphasis on security has restricted where research on certain biological agents may be conducted and who may participate in that research (DHHS, 2002).

ARE MEDICAL COUNTERMEASURES A PRIORITY?

Leaders in DoD and the nation have emphasized the seriousness of the biowarfare threat, but it is clear that they still have not made the development of additional FDA-licensed medical countermeasures to protect military troops a genuine priority. The committee sees dismal prospects for successful results (and no prospects for faster results) from the current efforts by DoD's Chemical and Biological Defense Program to produce medical countermeasures.

The successful development and licensure of new vaccines and drugs to protect against biological warfare agents cannot be done in half-measures. It requires instead a substantial and sustained commitment, including providing strong, scientifically knowledgeable leadership for an integrated and comprehensive effort supported by adequate funding. Differences between military and civilian circumstances in terms of perceived risks and medical strategies mean that despite a substantially ex-

TABLE 1-1 Budget Authority for the Medical Biodefense Component of the Chemical and Biological Defense Program (CBDP) and Other Selected DoD Research, Development, Testing, and Evaluation Programs, FY 1996–2004[a] (millions of dollars)

	Medical Biodefense (CBDP)[b]		
Fiscal Year	Science and Technology[c]	Advanced Development[d]	Total CBDP
1996	35.3	5.4	245.2
1997	33.7	13.1	283.1
1998	37.2	25.1	316.5
1999	41.4	23.1	320.2
2000	61.0	32.5	363.5
2001	64.8	44.2	377.6
2002	92.4	113.6	569.5
2003	104.5	84.2	598.1
2004[e]	92.8	73.9	559.6

[a]The data cover the budget categories of basic research (budget activity 6.1), applied research (6.2), advanced technology development (6.3), advanced component development and prototypes (6.4), and system development and demonstration (6.5).

[b]Includes activities related to vaccines, therapeutics, and diagnostics.

[c]Budget activities 6.1, 6.2, and 6.3.

[d]Budget activities 6.4 and 6.5.

[e]As requested in the President's FY 2004 budget.

NOTE: The data do not include funds for salary and benefits for military personnel participating in these activities or for the cost of facilities operated by the military services (e.g., USAMRIID). The scale of these additional costs is illustrated by data available for FY 2000 for USAMRIID, which accounts for the majority of military personnel involved in the medi-

panded civilian effort, unique military needs for medical countermeasures continue to exist, and meeting those needs is DoD's responsibility. Maintaining the status quo only assures a long, costly, and perhaps fruitless wait for new vaccines and therapeutic products.

The committee's concerns about DoD's management of its program to develop medical countermeasures against biowarfare agents should come as no surprise to DoD or Congress. Recent reports from a DoD-commissioned independent panel, IOM, the Defense Science Board, and RAND, as well as congressional testimony from the General Accounting Office, have reached many similar conclusions (Defense Science Board,

Biomedical Component of DARPA Biological Warfare Defense Program	Total DARPA Biological Warfare Defense Program	Missile Defense Agency
		2,886.4
		3,360.3
43.1	58.5	3,451.6
52.6	83.0	3,909.9
72.6	124.3	3,456.8
87.7	146.2	4,119.3
97.9	171.9	6,728.3
78.6	162.0	6,685.8
53.0	137.3	7,620.9

cal biodefense component of the Chemical and Biological Defense Program. Military personnel included 76 scientists and engineers and 134 technical support or other personnel. Military pay and allowances for these personnel for FY 2000 totaled $17.7 million. Operations and maintenance costs for USAMRIID for FY 2000 were $0.4 million. USAMRIID also had 113 civilian scientists and engineers and 120 civilian technical support personnel; the salaries and related costs for the civilian personnel engaged in medical biodefense research are included in the table. Funds for the separate DoD research and development program on infectious diseases are not included in this table. CBDP, Chemical and Biological Defense Program; DARPA, Defense Advanced Research Projects Agency.
SOURCES: DoD, no date, 1997–2003, 2002; Evans, 2003.

2002; GAO, 2002; IOM, 2002; Rettig and Brower, 2003; Top et al., 2000), including the following:

- the need for a single high-level authority within DoD that attends to the full spectrum of responsibility from threat definition through research and development, advanced product development, clinical trials, licensure, manufacture and procurement, and maintenance of manufacturing practice standards and regulatory compliance (Defense Science Board, 2002; IOM, 2002; Rettig and Brower, 2003; Top et al., 2000);
- the importance of having an ongoing, senior, external advisory group to maintain active relationships with current science and technol-

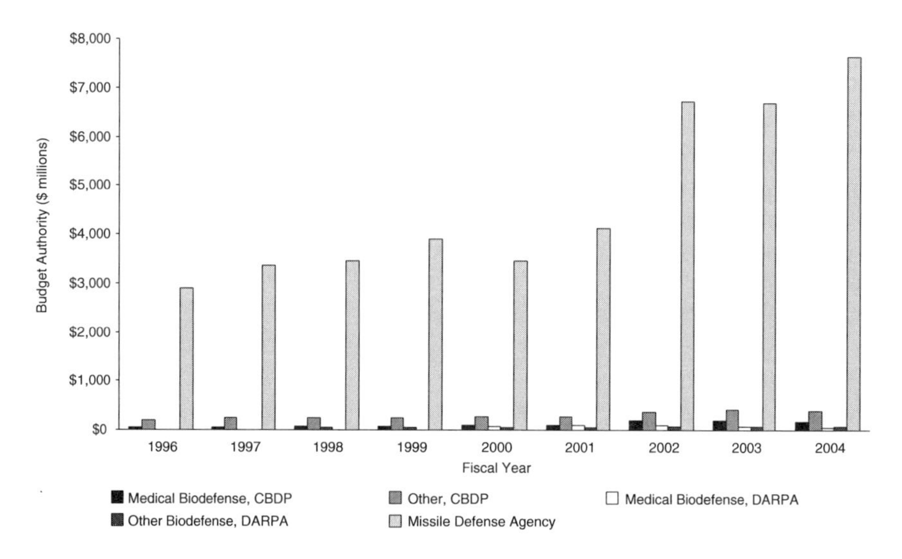

FIGURE 1-1 Budget authority for the medical biodefense component of the Chemical and Biological Defense Program and other selected DoD research, development, testing, and evaluation programs, FY 1996–2004. The medical biodefense component of the Chemical and Biological Defense Program includes activities related to vaccines, therapeutics, and diagnostics. The amounts for FY 2004 are the amounts requested in the President's FY 2004 budget. The data cover the budget categories of basic research (budget activity 6.1), applied research (6.2), advanced technology development (6.3), advanced component development and prototypes (6.4), and system development and demonstration (6.5). Excluded are amounts in the accounts of the military services for salary and benefits for military personnel participating in these activities and costs of facilities operated by the military services (e.g., USAMRIID). Funds for the separate DoD research and development program on infectious diseases are not included.
NOTE: CBDP, Chemical and Biological Defense Program; DARPA, Defense Advanced Research Projects Agency.
SOURCES: DoD, 1997–2003, 2002; Evans, 2003.

ogy leaders in the academic, corporate, and government sectors (IOM, 2002; Top et al., 2000); and

- the substantial investment required to successfully develop vaccines in particular, noting that DoD expenditures had not met those needs (IOM, 2002; Top et al., 2000).

Although changes have been made that appear to respond to some of these recommendations, the fundamental problems remain. Moreover, the committee believes that some of the recent actions will actually exacerbate the problems. In particular, in mid-2003 DoD implemented a recommendation from the Defense Science Board (2002) to assign responsibility for management of the science and technology component of the entire Chemical and Biological Defense Program to DTRA. No acceleration of medical biodefense science and technology should be anticipated from this change. Not only does DTRA have little experience or expertise in biomedical and pharmaceutical research, it adds yet another layer of management between key senior decision makers and the actual performance of research and development tasks for medical countermeasures.

RECOMMENDATIONS FOR CHANGE

A decision by DoD and national leaders to make DoD's program to develop medical countermeasures against biological warfare agents a genuine priority is the essential first step to set the stage for an effective program.

Thus, to ensure that DoD has an effective research and development program for medical biodefense countermeasures, the committee makes the following recommendation:

1. The Secretary of Defense and Congress must make the DoD program for medical biodefense countermeasures a high priority.

If that commitment is made, it has to be accompanied by major organizational and managerial changes:

- organizing the program to promote accountability and effective coordination throughout all phases of research, development, and product approval;
- installing leaders broadly knowledgeable in biotechnology with specific expertise in the development of vaccines and other pharmaceutical products;
- supporting the development of a strong scientific infrastructure, including scientific personnel with expertise in pharmaceutical product development and facilities for research and animal testing; and
- providing necessary funding to achieve program goals.

DoD should look to NIH as a model for promoting excellence in basic research and discovery science, and to the pharmaceutical industry for guidance in successful management of product development, a model that points to the importance of consolidating authority and responsibility. DoD's aim should be to have a program capable of meeting the need for not only vaccines and drugs based on conventional approaches, but also future products that emerge from today's exploration of new scientific concepts. An organizational approach that creates competing lines of authority and multiple reporting relationships, as the current matrix scheme does, is not adequate to address the multiple management and scientific challenges that DoD faces.

The program requires a single, highly knowledgeable leader who reports to a senior DoD policy official and whose sole responsibility is the direction of that program. The program director has to have budgetary control and full authority over—and accountability for—the entire range of program tasks, from basic research through product manufacturing. The program leadership and staff need to bring to the program substantial expertise in vaccine and pharmaceutical research, product development, and manufacturing, including the rapidly evolving contributions of biotechnology.

The organizational framework that shapes the planning, budgeting, management, and operation of the program has to be tailored to accommodate the following distinctive features of this work:

- For vaccines, the time from identification of a candidate product through FDA licensure is at least 7 to 12 years. It is too soon to know how long the development of new types of countermeasures (e.g., multivalent vaccines or immune modulators) will take, but important basic research remains to be done.
- The work with biological systems that vaccines require is inherently unpredictable.
- The risk of failure—which can result from problems with efficacy, safety, or manufacturing processes—remains high, even in relatively late stages of the development of pharmaceutical products.
- The overall cost of pharmaceutical product development varies, and much of the cost is incurred during the later stages of clinical testing.
- Vaccines and other pharmaceutical products must be tested and licensed in compliance with FDA oversight and regulation.
- FDA licensure of a vaccine is tied to a specific manufacturing facility.
- The need for human testing of vaccines and other pharmaceutical products demands attention to legal and ethical issues.

Many of these factors distinguish the development process for medical countermeasures from the engineering tasks that have driven the evolution of DoD's system for managing the development of new weapons systems and other products.

Furthermore, the present artificial separation between DoD's program to develop medical countermeasures against biological warfare agents and its parallel research and development program aimed at producing vaccines and other pharmaceutical products against infectious diseases of military significance should be eliminated. These two programs address similar scientific and technological questions and demand closely related expertise and facilities. Moreover, since concerns about biological warfare threats are expanding to include a wider range of naturally occurring and novel biological agents, the line between the two programs is becoming even less distinct and meaningful than it was in the past.

Also, Congress should seek guidance from DoD and others to assess the continuing appropriateness of its requirement that DoD devote close to 80 percent of its funding for medical biodefense research to work against biological agents that have been validated by intelligence assessments as near-term threats.[7] With a changing understanding of the range of potential threat agents and the ease with which some of them may be created, DoD may require greater flexibility in allocating funds between work on countermeasures against well-established threats and those that may not yet be validated.

In addition, DoD has to work with its counterparts in other government agencies to find ways to overcome other substantial obstacles to progress, especially the limited pool of specialized scientific expertise and research resources and the pharmaceutical industry's limited interest so far in producing these specialized products that have only a limited commercial market. Given that no federally owned facilities for full-scale manufacture of vaccines or drugs currently exist, industry must be considered an essential partner in efforts to make medical biodefense countermeasures available to DoD.

The committee considered a range of options for achieving these organizational, managerial, and scientific goals. After careful consideration and weighing the advantages and disadvantages of the various options, the committee recommends the creation of a new DoD agency:

2. Congress should authorize the creation of the Medical Biodefense Agency, a new DoD agency responsible for the research and development program for medical countermeasures against

[7]National Defense Authorization Act for Fiscal Year 1993, P.L. 102-484 (1992). A validated threat agent is defined in this legislation as one that is named in the biological warfare threat list published by the Defense Intelligence Agency (DIA) and is identified as a biowarfare threat by the Deputy Chief of Staff of the Army for Intelligence. A validated near-term threat is one that has been, or is being, developed or produced for weaponization within 5 years, as assessed and determined by the Defense Intelligence Agency (P.L. 102-484, Section 231).

biological warfare agents. The committee recommends the following features for this agency:

- **It should report directly to the Under Secretary of Defense for Acquisition, Technology, and Logistics.**
- **The functions of existing medical biodefense programs should be transferred to the new Medical Biodefense Agency, along with their personnel and funding, including the medical biodefense component of the Chemical and Biological Defense Program (including units within the Army such as USAMRIID) and related activities in DARPA. The research and development program for medical countermeasures against infectious diseases should also be transferred into the Medical Biodefense Agency.**
- **The Medical Biodefense Agency should function on the premise that to speed the development of countermeasures it is necessary to benefit fully from research and development efforts beyond DoD's intramural program so as to bring the expertise and creativity of industry and the academic community to the task.**
- **The Medical Biodefense Agency should ensure that DoD's medical biodefense activities are coordinated with and take full advantage of the related activities of NIH and that DoD's efforts are focused on meeting unique DoD needs.**

The proposal for the Medical Biodefense[8] Agency is discussed in detail in Chapter 2, as are alternative approaches considered by the committee. The committee considers the establishment of a newly designated agency as an essential step to accomplish program goals as well as to demonstrate the seriousness of DoD's commitment to its efforts to develop medical countermeasures against biological warfare agents. This action will help make the currently disjointed and poorly functioning program more effective by enhancing its stature and making the program leadership more directly accountable for performance.

Because the committee is strongly persuaded that creation of the Medical Biodefense Agency will be the most effective means of improving DoD's research and development program for medical biodefense countermeasures, much of the discussion and many of the recommendations throughout the report are framed in reference to this agency. In the event the Medical Biodefense Agency were not created, the need to establish a substantially more effective infrastructure for the DoD medical

[8] The term "biodefense" is used in this report to refer to defense against naturally occurring infectious diseases as well as biowarfare agents.

countermeasures program, as well as the need to address other critical challenges affecting that program, will remain and should be addressed by DoD. Those efforts can and should be guided by the same considerations that the committee discusses as the basis for its proposals for the Medical Biodefense Agency.

To promote accountability for action and progress by DoD and the Medical Biodefense Agency, an external review committee that is independent of DoD should also be established. The members of this committee should be experts drawn from academia, the biotechnology and pharmaceutical industries, and other segments of the private sector who are qualified to judge the plans and performance of the DoD research and development program for medical biodefense countermeasures.

> **3. Congress should establish an external review committee of experts in the development of vaccines and drugs to review and evaluate the program and performance of the DoD research and development program for medical biodefense countermeasures each year. The committee should report its findings each year to the Secretary of Defense and the Congress.**

The committee considers it highly preferable to maintain DoD control over the program to develop medical biodefense countermeasures to ensure that unique DoD needs are given high priority. However, DoD has failed to respond adequately to previous reports with similar recommendations for change. The committee believes that the development of biological warfare countermeasures requires the same urgency as work on medical bioterrorism countermeasures; therefore if DoD does not take steps sufficient to make the countermeasure development program effective, the committee recommends, as a last resort, transferring all or part of that responsibility from DoD to an agency responsible for promoting the development of medical countermeasures for bioterrorism defense. The external review committee should be charged with assessing progress after a 3-year period to recommend whether such a transfer should be made. The review committee must establish criteria for judging appropriate progress during this relatively brief interval, but it would be unreasonable to expect full resolution of all the problems currently affecting DoD's efforts to develop medical biodefense countermeasures.

If a transfer was considered appropriate, NIH appears, at present, to be the best alternative to a DoD-based program because of its depth of scientific expertise and its substantial funding to support work on medical defenses against bioterrorism. However, NIH has little tradition of product development or history of focus on military-specific needs, and among many competing national public health priorities this additional task may not be given sufficiently high priority.

4. If the review committee finds that after a 3-year period of operation the DoD research and development program for medical biodefense countermeasures has failed to make progress that the committee considers appropriate, Congress should transfer from DoD, in part or in whole, responsibility for the development of medical biodefense countermeasures and reassign that responsibility to an agency responsible for promoting the development of medical countermeasures for bioterrorism defense, such as the NIH or another agency considered appropriate.

The committee believes that the changes it is recommending for the development of medical biodefense countermeasures are consistent with the intent of DoD's own transformation initiative. The Secretary of Defense has described the war on terrorism as a "transformational event" that calls for rethinking current approaches to DoD's tasks (Rumsfeld, 2003). He sees a need for DoD to encourage a "culture of creativity and prudent risk-taking" to accomplish the transformation that he advocates for the department. Transformation is to extend beyond warfighting to the way in which the department is organized to support military personnel in the field (DoD, 2003b). It will require "commitment and attention from the [d]epartment's senior leadership" (DoD, 2003b, p. 3). The growth of asymmetric threats, including biological threats, is cited as one of the specific reasons for transformation. The strategies for accomplishing this transformation include (among others) encouraging innovative leadership and promoting rapid and innovative research and development programs.

Many issues must be addressed for DoD to have an effective research and development program for medical countermeasures against biological warfare agents. Those issues and the steps recommended by the committee are discussed in the remainder of this report. They are, however, secondary to the fundamental need for a strong and continuing commitment to the task.

REFERENCES

Bush GW. 2002. Remarks by the President at the signing of the Public Health Security and Bioterrorism Preparedness and Response Act of 2002, June.

Chu DS. 2002. Policy on Administrative Issues Related to Smallpox Vaccination Program (SVP). [Online]. Available: http://www.smallpox.mil/media/pdf/SPadminIssues policy.pdf [accessed November 18, 2003].

Clinton WJ. 1994. Executive order 12938—Proliferation of Weapons of Mass Destruction. *Weekly Compilation of Presidential Documents* 30(46):2386–2389. [Online]. Available: http://www.gpoaccess.gov/wcomp/search.html [accessed October 22, 2003].

Cohen WS. 1997. Message of the Secretary of Defense. In *Proliferation: Threat and Response*. Washington, DC: Department of Defense. [Online]. Available: http://www.defenselink.mil/pubs/prolif97/message.html [accessed September 24, 2003].

Cohen WS. 1998. Implementation of the Anthrax Vaccination Program for the Total Force. [Online]. Available: http://www.anthrax.osd.mil/media/pdf/implementationpolicy.pdf [accessed November 18, 2003].

Defense Science Board. 2001. 2000 Summer Study Executive Summary, Volume 1. In *Protecting the Homeland*. Washington, DC: Office of the Under Secretary of Defense for Acquisition, Technology, and Logistics.

Defense Science Board. 2002. *2001 Summer Study on Defense Science and Technology*. Washington, DC: Office of the Under Secretary of Defense for Acquisition, Technology, and Logistics.

DHHS (Department of Health and Human Services). No date. *FY 2004 Budget in Brief*. Washington, DC: Department of Health and Human Services. [Online]. Available: http://www.dhhs.gov/budget/04budget/fy2004bib.pdf [accessed March 5, 2003].

DHHS. 2002. Possession, use, and transfer of select agents and toxins; interim final rule. *Federal Register* 67(240):76886–76905.

DiMasi JA, Hansen RW, Grabowski HG. 2003. The price of innovation: new estimates of drug development costs. *Journal of Health Economics* 22(2):151–185.

DoD (Department of Defense). No date. *DoD In-House RDT&E Activities: FY2000 Management Analysis Report*. [Online]. Available: http://www.scitechweb.com/inhousereport/00cover.html [accessed September 9, 2003].

DoD. 1993. Department of Defense Directive: DoD Immunization Program for Biological Warfare Defense. Number 6205.3. Washington, DC: Department of Defense.

DoD. 1997–2003. Defense Budget Materials: Research, Development, Test & Evaluation Programs (R-1). [Online]. Available: http://www.defenselink.mil/comptroller/defbudget/fy2004/index.html [accessed February 3, 2003].

DoD. 2000. Department of Defense Directive: Use of Investigational New Drugs for Force Health Protection. Number 6200.2. Washington, DC: Department of Defense.

DoD. 2002. *Department of Defense Chemical and Biological Defense Program. Volume I: Annual Report to Congress*. Washington, DC: Department of Defense. [Online]. Available: http://www.acq.osd.mil/cp/nbc02/vol1-2002cbdpannualreport.pdf.

DoD. 2003a. *Department of Defense Chemical and Biological Defense Program. Volume I: Annual Report to Congress*. Washington, DC: Department of Defense. [Online]. Available: http://www.acq.osd.mil/cp/nbc03/vol1-2003cbdpannualreport.pdf [accessed June 27, 2003].

DoD. 2003b. *Transformation Planning Guidance*. Washington, DC: Department of Defense. [Online]. Available: http://www.oft.osd.mil [accessed August 1, 2003].

Evans D. 2003. Updated budget charts. E-mail to L Joellenbeck, Institute of Medicine, Washington, DC, June 24.

Fauci AS. 2003. NIAID biodefense research and collaborations with DoD. Presentation to the Institute of Medicine and National Research Council Committee on Accelerating the Research, Development, and Acquisition of Medical Countermeasures Against Biological Warfare Agents, Meeting III. Washington, DC.

FDA (Food and Drug Administration). 1995. Testing drugs in people. In *From Test Tube to Patient: New Drug Development in the United States, 2nd ed.* Rockville, MD: Food and Drug Administration. [Online]. Available: http://www.fda.gov/fdac/graphics/newdrugspecial/drugchart.pdf [accessed October 10, 2003].

FDA. 1999. *From Test Tube to Patient: Improving Health Through Human Drugs*. Rockville, MD: Food and Drug Administration, Center for Drug Evaluation and Research. [Online]. Available: http://www.fda.gov/cder/about/whatwedo/testtube-full.pdf [accessed July 25, 2003].

FDA. 2002. New drug and biological drug products: Evidence needed to demonstrate effectiveness of new drugs when human efficacy studies are not ethical or feasible. Final rule. 21 C.F.R. Parts 314 and 601. *Federal Register* 67(105):37988–37998.

GAO (General Accounting Office). 2002. *Chemical and Biological Defense: Observations on DoD's Risk Assessment of Defense Capabilities*. GAO-03-137T. Washington, DC: General Accounting Office.

Grant C. 2003. Aventis Pasteur Vaccine Development. Handout to the Institute of Medicine and National Research Council Committee on Accelerating the Research, Development, and Acquisition of Medical Countermeasures Against Biological Warfare Agents, Meeting III. Washington, DC.

IOM (Institute of Medicine). 2002. *Protecting Our Forces: Improving Vaccine Acquisition and Availability in the U.S. Military*. Lemon SM, Thaul S, Fisseha S, O'Maonaigh HC, eds. Washington, DC: The National Academies Press.

IOM. 2003. *Financing Vaccines in the 21st Century: Assuring Access and Availability*. Committee on the Evaluation of Vaccine Purchase Financing in the United States. Washington, DC: The National Academies Press.

Johnson-Winegar A. 2000. Department of Defense Anti-Biological Warfare Agent Vaccine Acquisition Program. Statement of Dr. Anna Johnson-Winegar, Deputy Assistant to the Secretary of Defense for Chemical/Biological Defense on April 14, 2000, before the Subcommittee on Personnel, Senate Armed Services Committee. [Online]. Available: http://armed-services.senate.gov/hearings/2000/p000414.htm [accessed July 3, 2003].

Linden C. 2002. Sponsor presentation on the study charge. Presentation to the Institute of Medicine and National Research Council Committee on Accelerating the Research, Development, and Acquisition of Medical Countermeasures Against Biological Warfare Agents, Meeting I. Washington, DC.

MacKenzie D. 2003. U.S. develops lethal new viruses. *New Scientist* 180(2419):6.

NRC (National Research Council). 2003. *International Perspectives: The Future of Nonhuman Primate Resources*. Washington, DC: The National Academies Press.

NVAC (National Vaccine Advisory Committee). 1999. Lessons learned from a review of the development of selected vaccines. *Pediatrics* 104(4):942–950.

Parker G. 2003. Biocontainment facilities needed to support medical biological defense research, development, test, and evaluation. Presentation to the Institute of Medicine and National Research Council Committee on Accelerating the Research, Development, and Acquisition of Medical Countermeasures Against Biological Warfare Agents, Meeting IV. Washington, DC.

Partnership for Public Service. 2003. *Homeland Insecurity: Building the Expertise to Defend America From Bioterrorism*. Washington, DC: Partnership for Public Service. [Online]. Available: http://www.ourpublicservice.org/publications3735/publications_show.htm?doc_id=181630 [accessed July 8, 2003].

Perry WJ. 1996. Preface: the new threat from nuclear, biological, and chemical weapons. In *Proliferation: Threat and Response*. Washington, DC: Department of Defense. [Online]. Available: http://handle.dtic.mil/100.2/ADA314341 [accessed September 24, 2003].

PhRMA (Pharmaceutical Research and Manufacturers of America). 2000. *Why Do Prescription Drugs Cost So Much and Other Questions About Your Medicines*. [Online]. Available: http://www.phrma.org/publications/publications/questions/questions.pdf [accessed July 2, 2003].

Rettig R, Brower J. 2003. The Acquisition of Drugs and Biologics for Chemical and Biological Warfare Defense: Department of Defense Interactions with the Food and Drug Administration. Santa Monica, CA: RAND Institute. [Online]. Available: http://www.rand.org/publications/MR/MR1659/[accessed October 20, 2003].

Public Citizen. 2001. *Rx R&D Myths: The Case Against the Drug Industry's R&D "Scare Card."* Washington, DC: Public Citizen's Congress Watch. [Online]. Available: http://www.citizen.org/publications/release.cfm?ID=7065&secID=1078&catID=126 [accessed June 23, 2003].

Rumsfeld D. 2003. Secretary's foreword. In *Transformation Planning Guidance.* Washington, DC: Department of Defense. [Online]. Available: http://www.oft.osd.mil [accessed August 1, 2003].

Struck MM. 1996. Vaccine R&D success rates and development times. *Nature Biotechnology* 14(5):591–593.

Top FH Jr., Dingerdissen JJ, Habig WH, Quinnan GV Jr., Wells RL. 2000. DoD Acquisition of Vaccine Production: Report to the Deputy Secretary of Defense by the Independent Panel of Experts. In DoD. 2001. *Report on Biological Warfare Defense Vaccine Research and Development Programs.* Washington, DC: Department of Defense. [Online]. Available: http://www.acq.osd.mil/cp/bwdvrdp-july01.pdf [accessed February 19, 2004].

U.S. Congress, House Armed Services Committee. 1993. *Countering the Chemical and Biological Weapons Threat in the Post-Soviet World.* Report of the Special Inquiry into the Chemical and Biological Threat. Committee print 15. 102nd Congress, 2nd Session. February 23.

U.S. Congress, Office of Technology Assessment (OTA). 1993. *Pharmaceutical R&D: Costs, Risks and Rewards.* OTA-H-522. Washington, DC: U.S. Government Printing Office. [Online]. Available: http://www.wws.princeton.edu/~ota/ns20/year_f.html [accessed July 8, 2003].

White House. 2003. President details Project BioShield. Washington, DC, February 3. [Online]. Available: http://www.whitehouse.gov/news/releases/2003/02/print/20030203.html [accessed February 5, 2003].

Wycoff RF. 1998. FDA's role in the medical, chemical, and biological warfare preparedness program of the Department of Defense. Testimony before the Senate Veterans' Affairs Committee. Washington, DC [Online]. Available: http://www.fda.gov/ola/1998/dod0398.html [accessed July 8, 2003].

2

Organizing Research and Development for Medical Countermeasures to Accomplish Essential Functions

The question posed to the committee was framed in terms of identifying opportunities for the Department of Defense (DoD) to accelerate the research, development, and acquisition of medical countermeasures against biological warfare agents. Drug and vaccine development is inherently challenging. As a result, the fundamental requirement for an effective program, and even more so for an accelerated program, is a firm and sustained commitment from DoD to make the task a high priority and to provide institutional and financial resources commensurate with the task.

In the committee's view, a necessary part of DoD's commitment is a reorganization of the department's efforts so as to maximize its opportunities and ability to capture the benefits of scientific advances in the discovery and development of medical countermeasures and to adapt the best practices of the pharmaceutical and biotechnology industries to its public sector system. This reorganization should promote a set of functional goals, reviewed below, that the committee considers essential. As outlined in Chapter 1, the recommended reorganization of DoD's efforts should be accomplished through creation of the Medical Biodefense Agency, a new agency in DoD that consolidates responsibility and authority over the entire research and development process for medical biodefense countermeasures. The committee believes that the prospect for timely advancement of medical countermeasures against all present and future biological warfare threats will be enhanced by the establishment of this agency, whose only agenda will be the development of such products. The Medical Biodefense Agency has to focus on products to meet the needs of military personnel, but it should also actively coordinate and collaborate with the related endeavors of the National Institutes of Health (NIH) to develop countermeasures for bioterrorism to benefit as much as possible from that work. The committee's proposal for this agency is described in more detail in this chapter. Other organizational options considered and rejected by the committee are also discussed.

A DOD AGENCY FOR ACQUISITION OF MEDICAL COUNTERMEASURES FOR BIOLOGICAL DEFENSE

As the result of a congressional mandate,[1] DoD's work on medical biodefense countermeasures is part of a program that addresses medical and nonmedical countermeasures against both chemical and biological warfare threats. Responsibility for centralized oversight of the Chemical and Biological Defense Program has been assigned to the Assistant to the Secretary of Defense for Nuclear and Chemical and Biological Defense Programs. However, the current operational reality is a fragmented process that puts research planning and activities for medical countermeasures under the direction of the Defense Threat Reduction Agency in the Office of the Secretary of Defense, while the execution of those activities (i.e., basic and applied research in a laboratory setting) rests largely with personnel of the U.S. Army Medical Research and Materiel Command (USAMRMC). Management of the acquisition process for candidate countermeasures that have reached the stage of advanced development is the responsibility of the Joint Program Executive Office for Chemical and Biological Defense, which operates under the direction of Army acquisition officials. The scientific and technical work of product development is being carried out by a variety of private sector firms and integrated through the prime systems contract with DynPort Vaccine Company (DVC). Program planning and budgeting are directed from within yet another DoD organization, the Joint Chiefs of Staff (see Figure 2-1).

To improve DoD's ability to effectively pursue the development and licensure of medical countermeasures against biological warfare agents, the committee strongly recommends the creation of the Medical Biodefense Agency, a new DoD agency with responsibility for all stages of the research and development process for these products. The committee's proposal for the Medical Biodefense Agency is guided by a set of functional goals, described in Box 2-1. By itself, creation of a new agency will not solve the problems that this committee and other observ-

[1]The National Defense Authorization Act for Fiscal Year 1994, P.L. 103-160 (1993).

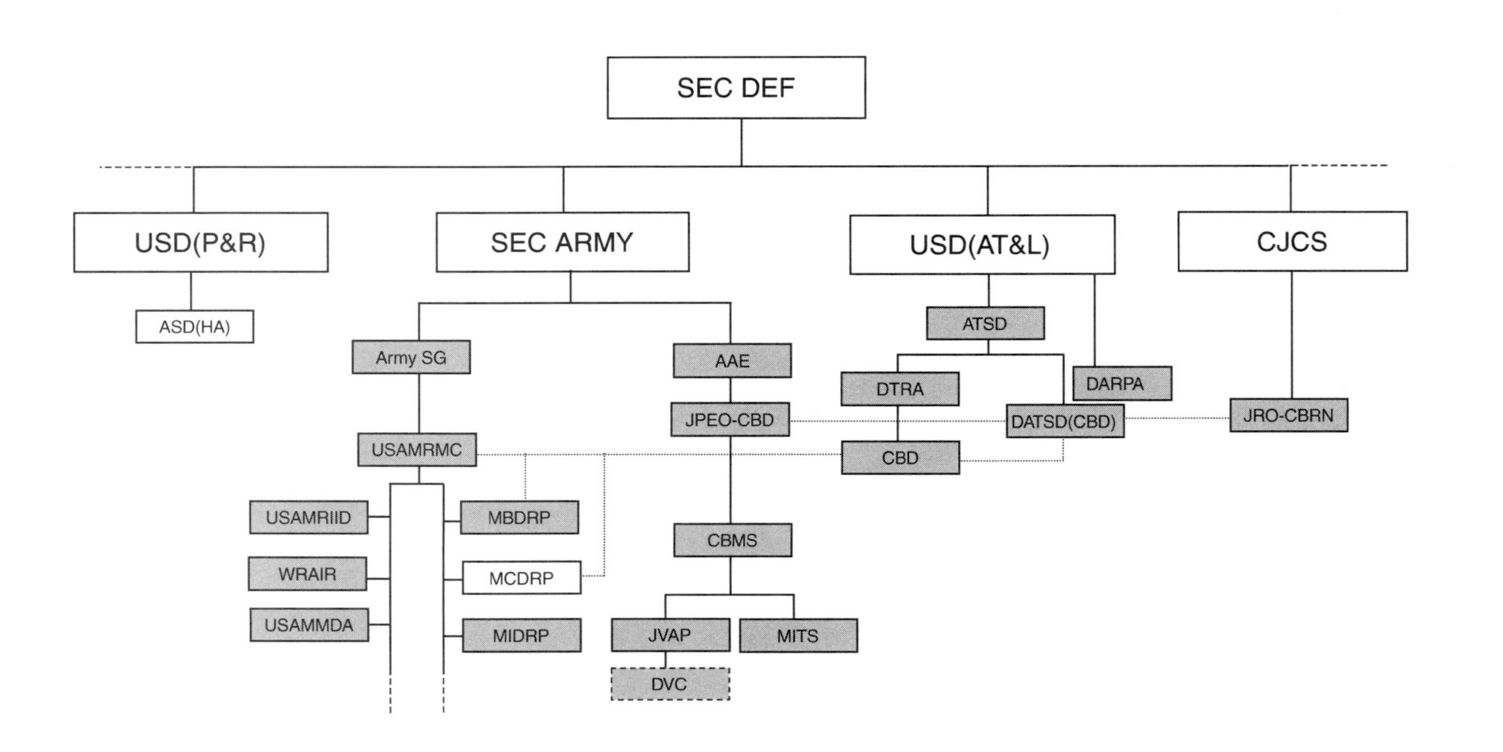
SEC DEF
USD(P&R)
ASD(HA)
SEC ARMY
Army SG
USAMRMC
USAMRIID
WRAIR
USAMMDA
MBDRP
MCDRP
MIDRP
AAE
JPEO-CBD
CBMS
JVAP
DVC
MITS
USD(AT&L)
ATSD
DTRA
CBD
DATSD(CBD)
DARPA
CJCS
JRO-CBRN

FIGURE 2-1 Simplified representation of current offices and organizations with a role in the DoD research, development, and acquisition program for medical biological defense (shaded boxes); dashed border indicates participant outside DoD. NOTE: **AAE**, Army Acquisition Executive; **Army SG**, Army Surgeon General; **ASD(HA),** Assistant Secretary of Defense for Health Affairs; **ATSD**, Assistant to the Secretary of Defense for Nuclear and Chemical and Biological Defense Programs; **CBD**, Directorate for Chemical and Biological Defense; **CBMS**, Chemical Biological Medical Systems; **CJCS,** Chairman of the Joint Chiefs of Staff; **DARPA**, Defense Advanced Research Projects Agency; **DATSD(CBD)**, Deputy Assistant to the Secretary of Defense (Chemical and Biological Defense); **DTRA**, Defense Threat Reduction Agency; **DVC**, DynPort Vaccine Company LLC; **JPEO-CBD**, Joint Program Executive Office for Chemical and Biological Defense; **JRO-CBRN**, Joint Requirements Office for Chemical, Biological, Radiological, and Nuclear Defense Programs; **JVAP**, Joint Vaccine Acquisition Program; **MBDRP**, Medical Biological Defense Research Program; **MCDRP**, Medical Chemical Defense Research Program; **MIDRP**, Military Infectious Diseases Research Program; **MITS**, Medical Identification and Treatment Systems; **SEC ARMY**, Secretary of the Army; **SEC DEF**, Secretary of Defense; **USAMMDA,** U.S. Army Medical Materiel Development Activity; **USAMRIID**, U.S. Army Medical Research Institute of Infectious Diseases; **USAMRMC**, U.S. Army Medical Research and Materiel Command; **USD**, Under Secretary of Defense: **(AT&L)** Acquisition, Technology, and Logistics**, (P&R)** Personnel and Readiness; **WRAIR**, Walter Reed Army Institute of Research.

BOX 2-1
Essential Functions of the DoD Research and Development Program for Medical Countermeasures Against Biological Warfare Agents

Establish and maintain knowledgeable leadership and effective management

A DoD program to develop vaccines and other pharmaceutical products should have strong, scientifically knowledgeable leaders at all levels. Together the program leadership and staff should bring expertise in areas ranging from basic research to animal and clinical testing to process development and product manufacturing (see Box 2-2). In addition, expertise in DoD acquisition and procurement is essential. The program also has to rely on others elsewhere in the government or in the academic community or industry to perform many scientific and technical tasks.

Offer effective identification, evaluation, and prediction of, as well as advocacy for, medical biodefense needs

Determining the need for medical countermeasures requires combining intelligence concerning biological threats, information on the characteristics of the threat agents, information on military planning for the use of medical countermeasures to maintain the effectiveness of forces on the battlefield, and the knowledge of the research and development community regarding medically and scientifically sound products. The focus should be on unique DoD needs or areas of expertise, considered within the context of work being planned or supported by other government agencies, the academic community, or the private sector.

Encourage and facilitate coordination with the related efforts of other government agencies, the academic community, and the private sector

DoD coordination with the Department of Health and Human Services and the Department of Homeland Security is essential to ensure that military service members benefit from the newer and better-funded drug and vaccine research and development efforts undertaken as part of the national defense against bioterrorism. Coordination across these agencies should also help DoD identify needs unique to biowarfare defense that are unlikely to be addressed without DoD action.

In addition, interagency cooperation should ensure that candidate products from DoD and unique DoD expertise and laboratory facilities are effectively used to support the nation's overall biodefense effort. A DoD program to develop vaccines and other biodefense countermeasures should also use the variety of mechanisms available, including grants, contracts, and cooperative research agreements, to cultivate partners in the academic community and the private sector through a strong and diverse set of extramural activities.

Seek necessary resources

An assumption of adequate financial resources underlies industry's estimate of a 7- to 12-year time frame for the development of pharmaceutical products. Accelerating the process is likely to require higher funding levels to speed work on particular candidate products and increase the number of candidates that can move through the research and development process at the same time. In addition, increased funding will be needed to allow for an increased risk of failure in efforts to develop candidate products in an accelerated program. With limited funding at present, the pace of product development is likely to be set by budget constraints rather than by scientific opportunity and product readiness.

In addition to appropriate funding, the program requires access to other resources in the form of research and product development infrastructure. Key infrastructure components include a highly skilled workforce, appropriate types and numbers of research animals, and specialized laboratory and production facilities. Many elements of this infrastructure are currently in short supply, not only for DoD but for the nation as a whole.

Promote program stability

The time required to bring vaccines and other pharmaceutical products from discovery to licensure by the Food and Drug Administration (FDA) demonstrates the need for continuity of effort and commitment, assumption of risk, and acceptance of a certain degree of failure. For DoD to contemplate meeting or improving upon industry's cycle time, disruptions to the planning and execution of the research and development program must be minimized. Both budgetary and program planning and execution should also allow for prompt responses to the likely occurrence of unforeseeable problems that can arise at any stage of product development.

Understand and promote the use of the best science for the task

Because the discovery and successful development of medical biodefense countermeasures is a challenging and uncertain task requiring substantial and varied expertise in rapidly changing fields, as indicated above, DoD should support both intramural and extramural work. For its intramural activities, DoD should have the means to attract strong scientific and technical talent on a permanent or temporary basis. An environment supportive of scientific exchange and rigor is needed to foster and maintain excellence in basic research and the product development pro-

Continued

BOX 2-1 Continued

cess. Mechanisms for obtaining independent, expert advice for overall program planning and for peer review of specific research proposals and program activities are also essential.

Tailor the acquisition process for medical countermeasures to use only FDA's regulatory requirements as the basis for assessing the technical merits of candidate products

For vaccines and other pharmaceutical products, including those intended for use as medical biodefense countermeasures, product testing and licensure are regulated by the FDA. DoD product development programs, however, are managed through the defense acquisition system, which is oriented to engineering methods and standards primarily for development and testing of mechanical and electronic equipment for weapons systems or software and equipment for information systems. For DoD to work effectively toward licensure of medical countermeasures, the acquisition process must be tailored to the requirements of FDA regulatory oversight, a step endorsed in principle in current DoD policy (DoD, 2003). Establishing the efficacy of biodefense countermeasures, which requires relying on evidence from animal studies, will pose unfamiliar challenges for both product developers and FDA. These new challenges make it especially important that the countermeasure development process in DoD rest on a strong base of both scientific expertise and knowledge of FDA regulatory standards and requirements.

Provide the means for obtaining expert advice on ethical and legal issues

A DoD program to develop medical biodefense countermeasures requires access to expert advice on the ethical and legal issues raised by the testing and use of these products. Although these countermeasures are being developed to protect military personnel, the clinical testing now under way is relying on civilian volunteers. Including military personnel in these studies poses the challenge of establishing truly voluntary participation in a military setting. However, depending solely on civilian volunteers who are unlikely to ever benefit from their participation and who may be responding to economic incentives poses ethical questions as well (Tishler and Bartholomae, 2002). Well-informed ethical and legal advice will also be essential should consideration be given to emergency use of medical countermeasures not yet licensed by FDA.

BOX 2-2
Intramural Expertise Needed for a DoD Research and Development Program for Vaccines and Other Medical Countermeasures

- Defining product development priorities using information from intelligence sources on biological threats
- Basic research to identify candidate countermeasures against specific biological agents
- Animal models
- Process development and product development
- Pilot lot manufacturing
- Surrogate endpoint assessment in animal models and clinical trials
- Design and execution of clinical testing for Phase 1, Phase 2, and Phase 3 trials of safety and efficacy
- Analytical testing and quality control of candidate products
- Preparation of testing and consistency lots in compliance with FDA requirements for Good Manufacturing Practice
- Preparation and submission of product testing and licensing applications to FDA
- Full-scale manufacturing
- Contracting and business strategy

ers (Defense Science Board, 2002; IOM, 2002; Top et al., 2000) have identified. There must also be a genuine commitment to the development of medical biodefense countermeasures from senior DoD leadership, a commitment that can be demonstrated by providing the new agency with outstanding leadership, supporting scientific excellence, and providing necessary resources.

Key Requirements for the Medical Biodefense Agency

The committee's recommendation calls for Congress to authorize the creation of the Medical Biodefense Agency to be responsible specifically and exclusively for all aspects of DoD's research and development program for medical countermeasures against biological warfare agents and infectious diseases of military significance. This agency and its director should report directly to a Senate-confirmed position in the Office of the Secretary of Defense, preferably the Under Secretary of Defense for Acquisition, Technology, and Logistics (see Figure 2-2).

The functions of a variety of existing DoD organizations and activities should be brought together under the control of the Medical Biodefense

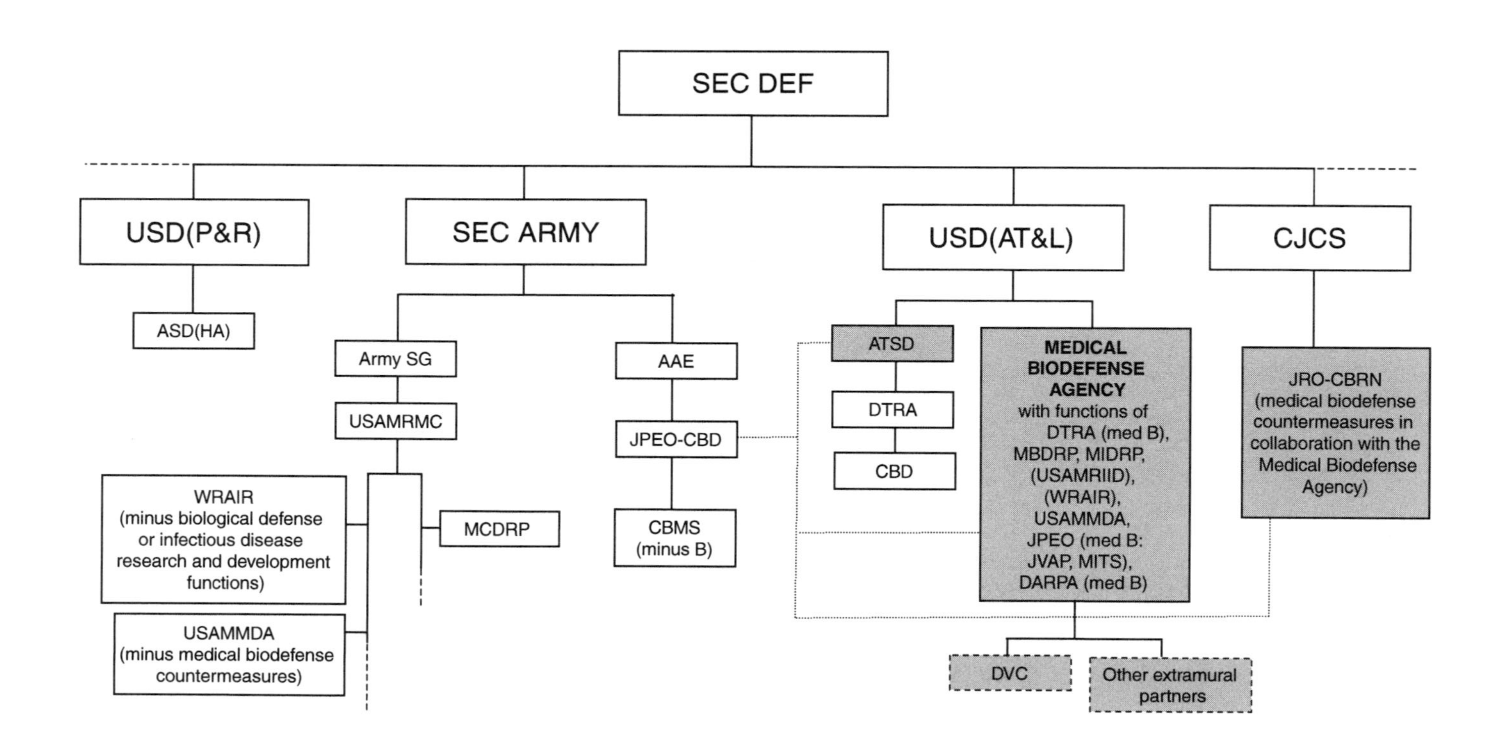

SEC DEF
USD(P&R)
ASD(HA)
SEC ARMY
Army SG
USAMRMC
WRAIR
(minus biological defense or infectious disease research and development functions)
USAMMDA
(minus medical biodefense countermeasures)
MCDRP
AAE
JPEO-CBD
CBMS
(minus B)
USD(AT&L)
ATSD
DTRA
CBD
MEDICAL BIODEFENSE AGENCY
with functions of DTRA (med B), MBDRP, MIDRP, (USAMRIID), (WRAIR), USAMMDA, JPEO (med B: JVAP, MITS), DARPA (med B)
DVC
Other extramural partners
CJCS
JRO-CBRN
(medical biodefense countermeasures in collaboration with the Medical Biodefense Agency)

FIGURE 2-2 Placement of the proposed Medical Biodefense Agency within DoD. Shaded boxes indicate components with a role in the DoD research, development, and acquisition program for medical biological defense; dashed border indicates participant outside DoD. Please see text for explanation of proposed roles and responsibilities. NOTE: **AAE**, Army Acquisition Executive; **Army SG**, Army Surgeon General; **ASD(HA)**, Assistant Secretary of Defense for Health Affairs; **ATSD**, Assistant to the Secretary of Defense for Nuclear and Chemical and Biological Defense Programs; **CBD**, Directorate for Chemical and Biological Defense; **CBMS**, Chemical Biological Medical Systems (**minus B**: without the medical biodefense functions); **CJCS**, Chairman of the Joint Chiefs of Staff; **DARPA (med B),** the functions of the Defense Advanced Research Projects Agency for medical countermeasures against biowarfare agents; **DTRA**, Defense Threat Reduction Agency; **DTRA (med B),** functions currently carried out by DTRA for medical countermeasures against biowarfare agents; **DVC**, DynPort Vaccine Company LLC; **JPEO-CBD**, Joint Program Executive Office for Chemical and Biological Defense; **JPEO (med B),** the functions of the Joint Program Executive Office for Chemical and Biological Defense for medical countermeasures against biowarfare agents; **JRO-CBRN**, Joint Requirements Office for Chemical, Biological, Radiological, and Nuclear Defense; **MBDRP**, Medical Biological Defense Research Program; **MCDRP**, Medical Chemical Defense Research Program, **MIDRP**, Military Infectious Diseases Research Program; **MITS**, Medical Identification and Treatment Systems; **SEC ARMY**, Secretary of the Army; **SEC DEF**, Secretary of Defense; **USAMMDA**, U.S. Army Medical Materiel Development Activity; **USAMRIID**, U.S. Army Medical Research Institute of Infectious Diseases; **USAMRMC**, U.S. Army Medical Research and Materiel Command; **USD**, Under Secretary of Defense: **(AT&L)** Acquisition, Technology, and Logistics, **(P&R**) Personnel and Readiness; **WRAIR**, Walter Reed Army Institute of Research.

Agency. This includes the responsibilities related to the development of medical countermeasures against biowarfare agents that currently lie within the following organizations:

- Medical Chemical and Biological Defense Directorate, Defense Threat Reduction Agency;
- Research Area Directorate for Chemical and Biological Defense, USAMRMC;
- U.S. Army Medical Research Institute of Infectious Diseases, USAMRMC; and
- Chemical Biological Medical Systems, Joint Program Executive Office for Chemical and Biological Defense.

The new agency should also assume responsibilities related to the development of infectious disease countermeasures in the following organizations:

- Research Area Directorate for Infectious Diseases, USAMRMC;
- Pharmaceutical Systems Project Management Division, U.S. Army Medical Materiel Development Activity; and
- Components of Walter Reed Army Institute of Research, USAMRMC.

Other DoD activities related to research and development of pharmaceutical products should be reviewed to determine whether it would be appropriate to incorporate them in the new agency as well.

The Medical Biodefense Agency should be expected to coordinate with the remaining functions of the Chemical and Biological Defense Program (medical and nonmedical countermeasures against chemical warfare threats, and nonmedical countermeasures against biological threats) through the Assistant to the Secretary of Defense for Nuclear and Chemical and Biological Defense Programs, who also reports to the Under Secretary of Defense for Acquisition, Technology, and Logistics.

The reasons for this recommendation and specific features of its implementation are discussed here.

Creation of a New DoD Agency

DoD's efforts to develop medical biodefense[2] countermeasures have suffered because problems with the organization and management of

[2]Throughout the remainder of this report the term "biodefense" is used to describe defense against naturally occurring infectious diseases as well as biowarfare agents.

medical countermeasures activities have not been addressed. Further, there is no indication that senior DoD leaders have given the development of these products the priority required for success.

The creation of a newly designated agency reporting to an Undersecretary within the Office of the Secretary of Defense is a major step. The committee considers such a step necessary for two reasons: to integrate all elements necessary to support accomplishing the task of developing medical biodefense countermeasures, and to centralize responsibility and authority for this task. Creating the new agency and placing it at a high level within DoD is intended to bring visibility and accountability to a program that is functioning poorly. Recommending that the Medical Biodefense Agency report directly to an undersecretary is also intended to indicate a high priority for these activities within DoD and to improve the agency's ability to advocate for appropriate funding levels. Furthermore, establishing the agency at this level should enhance the status of the program and help DoD attract highly qualified leaders who can improve its program by applying expertise from the successful development of other vaccines and drugs.

Creation of the Medical Biodefense Agency will establish a more appropriate framework for managing and conducting the research and development that should lead to medical countermeasures licensed by FDA. The research and development process for vaccines and other pharmaceuticals is most effective when it is based on early and continuing collaboration among laboratory scientists, manufacturing engineers, clinical investigators, and regulatory affairs specialists. The committee's proposal would remedy the current organizational divide between laboratory science and subsequent stages of product development by placing all of those activities under the direct authority of the head of the new agency.

The organizational aspects of the committee's recommendation for the Medical Biodefense Agency reflect the conclusion that a clear break with past approaches is necessary. Before deciding to recommend that the Medical Biodefense Agency report to the Under Secretary of Defense for Acquisition, Technology, and Logistics, the committee considered the merits of placing the program under the Assistant Secretary of Defense for Health Affairs because of the link that could be provided to military medical expertise. However, the preponderant focus of the Assistant Secretary's office on medical care and health insurance was viewed as having insufficient relevance to a research and development program for vaccines and drugs for biowarfare defense. In contrast, the Under Secretary of Defense for Acquisition, Technology, and Logistics is specifically responsible for overseeing the department's various research and development activities and the subsequent work necessary to bring to a finished state products developed to meet DoD needs. The recommendation for the Medical Biodefense Agency is aimed at creating a framework for such activity that is appropriate for pharmaceutical products.

The committee also concluded that placing the new agency within an existing organization reporting to the Under Secretary, or perhaps within one of the Army components that has had a major role in medical countermeasures work in the past, offered little prospect of indicating a true change in priority for the program or improving its visibility and prospects for appropriate funding. Furthermore, attempting to implement a new and atypical approach for managing research, development, and acquisition within an existing organization that is otherwise structured to follow current practices seems to promise little support for the changes the committee considers essential.

Scope of Responsibility

The Medical Biodefense Agency should have responsibility for all aspects of DoD-sponsored research and development for medical biodefense countermeasures. The products that would fall under the purview of this agency include vaccines, therapeutic drugs, antitoxins, and diagnostics directed against biological agents. All of these products require FDA licensure for routine use.

It is essential that the Medical Biodefense Agency have responsibility and authority for developing program plans and budgets. The programmatic tasks to be overseen or conducted should include basic research through advanced development. Thus, the agency would be responsible for activities that DoD defines as "science and technology" and for the acquisition programs for candidate products that move into advanced development. All of these activities should be under the new agency's direct control. The existing units currently carrying out these tasks would be absorbed into the new agency and reconfigured to create as efficient an organization as possible. The existing matrix organizational scheme that creates competing lines of responsibility and authority is not adequate.

Of particular importance is ensuring that the Medical Biodefense Agency has the authority to manage the transition of candidate products from the science and technology stage into, and their progress through, the DoD acquisition system. At present, this transition occurs before clinical testing begins. Formal entry of a candidate product into the acquisition system superimposes DoD reporting and review requirements on the already stringent documentation and reporting requirements associated with the FDA's regulation of clinical testing and product development in support of an application for licensure of a medical countermeasure.

Premature transfer of candidate products into the acquisition system was cited in discussions with the committee as a factor in delays. In reviewing DoD technology development programs more generally, the General Accounting Office (GAO, 1999) recommended additional development of products within the science and technology framework but noted that necessary funding for such work is often more readily available only when a project has entered the acquisition system. To improve the development process for medical countermeasures, the Medical Biodefense Agency should have the authority to use funds from science and technology accounts (e.g., budget activity 6.3) to support Phase 1 and even Phase 2 clinical trials before a candidate product is subject to acquisition system review. In addition, the agency director requires the authority to allocate funding across budget activity categories in a manner that provides appropriate resources for the tasks deemed appropriate for each category.

Because the committee expects the Medical Biodefense Agency to promote the highest scientific standards, the agency's work would be conducted through a mix of intramural and extramural activities. Thus, the agency needs the authority to use grants, contracts, and a full range of other acquisition mechanisms to engage extramural partners. The agency also needs the authority to create training programs to help prepare new investigators and product developers. Mechanisms for meeting these needs are discussed in Chapter 3.

Management of DoD's prime systems contract for the development, licensure, and initial manufacture of several biodefense vaccines should become the responsibility of the new agency. In addition, the new agency should have the option to develop partnerships with other drug and vaccine companies or other entities, including academia, to carry medical countermeasures through clinical trials, scale-up, and licensure. However, given that vaccine development under the prime systems contract is still in the early stages, the committee believes that DoD must consider the possibility that private sector firms may ultimately be reluctant or unable to supply the manufacturing capacity necessary to meet its needs. Because the manufacturing facility is integral to the licensure of a vaccine, the Medical Biodefense Agency should have the authority and responsibility to explore a full range of options for vaccine production, including government ownership or operation of such a facility (see DoD, 2001, 2002).

The committee also recommends incorporating into the Medical Biodefense Agency's portfolio the DoD research and development program for vaccines and drugs to protect against infectious diseases of military importance. The infectious disease program is closely related to the medical biodefense program in terms of the scientific and technological challenges addressed. Bringing the two programs together would enable each of them to benefit from access to a broader range of intramural and extramural expertise. In addition, the prospect of commercial markets for at least some products from the infectious disease program has led to industry partnerships that might prove helpful in the development of biodefense products.

Leadership and Staffing

It is essential that the Medical Biodefense Agency have a highly qualified director who has strong experience in vaccine and drug research and development. The position can be open to either a civilian or military appointment, but it should be filled at the flag, Senior Executive Service, or equivalent level on the basis of appropriate qualifications. For example, training in clinical medicine without experience in laboratory research or product development and testing would not be adequate. The agency should also have strong medical expertise in infectious diseases at senior management levels. Because vaccines and drugs are based on different scientific principles and require different approaches for development and production, it may be appropriate for the agency to have a senior deputy for each discipline.

Every effort should also be made to fill managerial positions throughout the agency with persons having appropriate types and levels of expertise (see Box 2-2). Even though the Medical Biodefense Agency will rely on extramural partners to perform many tasks, it is essential that the agency have a critical core of expertise available within its own staff. In addition to scientific and product development expertise, the skills of the agency staff have to include DoD acquisition and procurement policies and procedures. Chapter 3 discusses mechanisms that the Medical Biodefense Agency might employ to attract the highly skilled staff it requires.

It is crucial that the director have authority over the agency's budget and its full range of management and operational activities and be the advocate within DoD for appropriate funding and programmatic priority. In addition, the director should be the agent of coordination for the Medical Biodefense Agency within DoD and the military services. The Medical Biodefense Agency has to be effectively represented within the high-level decision-making processes that set DoD and national priorities, and the agency director should be DoD's primary representative to other government agencies on matters related to the development of medical countermeasures for defense against biological agents. In particular, the director of the Medical Biodefense Agency should be the principal DoD representative to the Interagency Working Group on Weapons of Mass Destruction Medical Countermeasures, chartered by the National Science and Technology Council to coordinate the efforts and responsibilities of federal agencies (Interagency Working Group on Weapons of Mass Destruction Medical Countermeasures, 2003). This responsibility will require that the agency and its director be well versed in medical biodefense issues related to both biowarfare and bioterrorism. Coordination with other national efforts should ensure that DoD benefits maximally from research and development efforts directed toward bioterrorism countermeasures, while being able to focus its own efforts on the military's unique needs.

Program Priorities and Planning

The Medical Biodefense Agency should have the authority to determine its program priorities and program plans in collaboration with the Joint Requirements Office for Chemical, Biological, Radiological, and Nuclear Defense.

Requirements for new products to meet military needs, including medical countermeasures, are normally established by the military services through the Joint Chiefs of Staff. The committee is concerned that the current process incorporates too little expertise in vaccine and drug research and product development to assign the most scientifically appropriate priorities and to develop program plans and budgets for medical biodefense countermeasures. The newly established Joint Requirements Office for Chemical, Biological, Radiological, and Nuclear Defense (chartered in February 2003) promises to give new visibility and coherence to the requirements-setting process for the Chemical and Biological Defense Program as a whole. However, the organization's charter indicates only a single assigned medical materiel requirements position (Pace, 2003).

For the Medical Biodefense Agency to be able to set priorities appropriately, it must have access to information on biological threats and military force planning, as well as on the efforts undertaken or planned by NIH's National Institute of Allergy and Infectious Diseases (NIAID) or other federal agencies that might also help in meeting military needs. The information on the risk that various biological agents are thought to pose must be considered along with informed judgments of the scientific feasibility of countermeasure development and the pace at which development can be expected to proceed.

Budget

The committee believes that for its current scope the DoD program to develop medical biodefense countermeasures is underfunded based on the experience of other relevant government agencies and the private sector. The committee also believes that the program should be better focused before any substantial increase in funding occurs. The current program, supporting research and development activities on products against more than 10 biologic agents—with 7 candidate products now in clinical testing—is ambitious by industry standards (Top et al., 2000) and too diffuse to be effective with the resources available. In addition, development costs can vary widely among pharmaceutical products and generally increase substantially as clinical testing progresses, making it necessary to consider the specific mix of projects to make a realistic estimate of annual funding needs.

The committee advises that the agency's budget should initially be based on the funding currently allocated to the research and development activities for medical biological defense in the Chemical and Biological Defense Program ($189 million for fiscal year [FY] 2003 in budget activities 6.1 through 6.5), funding for related activities currently supported by the Defense Advanced Research Projects Agency (DARPA) ($79 million for FY 2003), and funding for the research and development program for vaccines and drugs against infectious diseases ($54 million for FY 2003). Related management support funding for each of these program areas (budget activity 6.6) should also be included.

In addition to this baseline of $322 million, the initial budget should include an increase of $100 million the first year, rising over the first 5 years to $300 million above the baseline amount. This increase reflects, in part, the expectation that more work will be done by civilian instead of military personnel and in non-DoD facilities, via a vibrant extramural program. The costs of salaries for military personnel and the operation of military facilities such as the U.S. Army Medical Research Institute of Infectious Diseases (USAMRIID) do not appear in budgets for the Chemical and Biological Defense Program or the infectious disease program since they are presently covered in accounts of the military services. In addition, some candidate products are moving into later, more costly phases of development. DoD and Congress should expect the new agency's funding needs to increase further as more products reach this stage. Funds for renovating or replacing USAMRIID facilities should also be added to the Medical Biodefense Agency's budget. The committee emphasizes that once countermeasures are licensed, additional funding beyond this research and development budget will be needed for procurement of operational quantities of those products.

The committee did not use the average cost to the sponsor of developing a new vaccine or drug in formulating its advice regarding the agency's budget. Controversy has surrounded the estimates that have been made over the past 25 years (U.S. Congress, OTA, 1993).[3] As noted in Chapter 1,

---

[3]Despite these controversies, widespread consensus exists that the full cost to a sponsor of research and development, from discovery through clinical testing to FDA approval, should include the following components: (1) the amount spent for drugs (or vaccines) that success-

the congressional Office of Technology Assessment (OTA) in 1993 estimated the average cost of bringing a new drug to licensure to be $237 million in 1990 dollars (U.S. Congress, OTA, 1993), and more recent estimates have ranged from $110 million to $802 million (2000 dollars) (DiMasi et al., 2003; Public Citizen, 2001). Concerns about the most widely quoted estimates have centered on the representativeness of the companies and drugs on which researchers based the estimates, the accuracy of the unaudited costs reported, and the appropriateness of the methods used (e.g., whether to report pre-tax or after-tax costs) (DiMasi et al., 2003, Public Citizen 2003, U.S. Congress, OTA, 1993). For the present purpose, those estimates may also be somewhat unsatisfactory because they are derived primarily from data for new drugs rather than vaccines, which are the principal focus of DoD's product development efforts.

The director of the Medical Biodefense Agency should be expected to review all current programs and to develop a strategic plan that will become the basis for appropriate budget requests in the future. As noted above, DoD and Congress should expect agency requests to increase over current levels as candidate vaccines and other products enter and move through the clinical testing process, traditionally the most costly stage of product development. For vaccines, another factor that may contribute to additional costs as the product development process moves forward is the need for dedicated manufacturing facilities (either small-scale or full production capacity) that can meet FDA requirements for compliance with current Good Manufacturing Practice (cGMP).[4]

External Scientific Advice and Oversight

An External Review Committee A legislatively mandated external review committee should be established to provide input to the director of the Medical Biodefense Agency and to monitor progress in improving the effectiveness of DoD's efforts to develop medical biodefense countermeasures. The members of the review committee should be experts in various aspects of vaccine and drug research and development, drawn from academia, industry, and other relevant organizations in the private

---

fully received approval for marketing; (2) the amount spent for products that failed at some point in the process; (3) the timing of expenditures on these costs; and (4) the cost of investing capital in this research and development process instead of alternative uses.

[4]Current Good Manufacturing Practice in Manufacturing, Processing, Packing, or Holding of Drugs; General (21 C.F.R. Part 210 [2003]); Current Good Manufacturing Practice for Finished Pharmaceuticals (21 C.F.R. Part 211 [2003]); and Biological Product Standards (21 C.F.R. Parts 600-680 [2003]).

sector (e.g., foundations). In consultation with the agency director, the review committee should develop a set of product-oriented performance criteria that can be used to evaluate the Medical Biodefense Agency. The review committee should meet at least twice a year to ensure that members gain a good understanding of both the scientific and operational aspects of the agency's work. The committee should submit an annual evaluation of the agency's performance to the Secretary of Defense and the Congress. As discussed in Chapter 1, the review committee should also be charged with assessing the agency's progress after 3 years and recommending whether to continue with a DoD-based program to develop medical countermeasures.

These review responsibilities should not be assigned to existing advisory groups such as the Armed Forces Epidemiology Board, the Defense Science Board, or the Defense Threat Reduction Agency's Threat Reduction Advisory Council. Each of these groups must address a broad range of subjects, and they lack the specialized expertise in the development of pharmaceutical products that the committee considers necessary for this task.

Scientific Advice and Peer Review Mechanisms for obtaining independent, expert advice for overall program planning and for assessments of specific research proposals and program activities are also essential. The groups of scientists convened to conduct peer reviews of proposals submitted to NIH (the NIH "study sections") and the advisory committees for individual NIH institutes or for the Centers for Biologics and Drug Evaluation and Research at FDA are ready models for DoD to use to guide funding decisions about both intramural and extramural programs. By calling on scientists from the broader research community, NIH and FDA gain the benefit of additional expertise and independent judgments of scientific merit and program priorities. For the Medical Biodefense Agency, this type of collaboration with extramural scientists can be a source not only of guidance on program priorities and funding decisions, but also of additional scientific advice for intramural scientists and program managers. In addition, it can promote wider knowledge of and interest in the DoD research and development program. The periodic program area reviews commissioned from the American Institute of Biological Sciences (e.g., AIBS, 2001), while appropriate, do not offer the kind of prospective guidance or ongoing interaction with the larger scientific community that the committee considers essential. The Scientific Steering Committees currently used by DoD for programmatic development and review do not take sufficient advantage of experts beyond DoD and other government agencies.

Incorporating New Science for the Longer Term

Much of the work in a DoD program to develop medical biodefense countermeasures must be focused clearly on bringing specific products to licensure. Yet new scientific and technical pathways are being opened up by rapid advances in biotechnology. A means should exist for a DoD countermeasures program to support a limited amount of highly innovative work that may prove useful only in the longer term. DARPA might serve as a model for this type of research support. DARPA program managers define topic areas of interest and use a variety of mechanisms (e.g., grants, contracts, SBIR[Small Business Innovation Research]/STTR [Small Business Technology Transfer] awards, and other transactions) to direct funding for research projects to university and industry scientists. To encourage fresh views on the most promising research needs or opportunities, DARPA program managers are recruited from outside DoD and usually serve for terms of only 4 years. The projects on medical biodefense countermeasures that DARPA itself has funded since FY 1998 will soon be ending. Incorporating a DARPA-like function within the Medical Biodefense Agency would help ensure continuing access to innovative work in the rapidly evolving fields of vaccine and drug research.

Potential Shortcomings of a Medical Biodefense Agency

The committee recognizes the challenges posed by its recommendation that a new DoD agency be created to conduct all aspects of the department's effort to develop medical countermeasures against biological warfare agents. Organizational change is disruptive, and substantial "cultural" differences among existing units must be reconciled to create a cohesive agency. The effectiveness of the Medical Biodefense Agency will also depend, in part, on its receiving strong support from senior DoD leaders, something the medical countermeasures effort does not appear to have had in the past.

The new agency has to attract a highly qualified director and bring to the management of the program substantially more expertise in the development of vaccines and other pharmaceutical products, despite the widespread shortage of expertise in fields related to biodefense (Partnership for Public Service, 2003). DoD may have difficulty competing for the talent it needs with NIH and other government agencies, academia, and industry.

Under the committee's recommendation, plans for advanced development of medical countermeasures, leading to FDA licensure and full-scale production, continue to rest primarily on contracting for services with companies in the private sector. In general, however, there have been barriers to pharmaceutical firms' participation in this work, which are described further in Chapter 3. DoD has sought to overcome these barriers through the use of a prime systems contractor, DVC, which is managing and integrating work being done by as many as two dozen other companies (DVC, 2003). However, DVC has yet to demonstrate that its "virtual company" can succeed in licensing a product.

CONSIDERATION OF ALTERNATIVE APPROACHES

After considering other alternatives, the committee concluded that it is necessary to create a DoD-based Medical Biodefense Agency for the development of medical biodefense countermeasures to ensure that unique DoD needs are addressed. The committee weighed and firmly rejected as inadequate the option of making minor adjustments within the existing DoD organizational framework or consolidating the research and development activities for medical countermeasures under existing organizations within DoD. Even if circumstances were to make alternatives such as these appealing to policymakers, it would remain necessary to find solutions to the same organizational and managerial problems, as well as the broader challenges discussed in Chapter 3, that led the committee to its recommendation to create the Medical Biodefense Agency.

Other alternatives considered by the committee looked beyond DoD. Because of the substantial resources that have recently been provided to NIAID to address the research and development of bioterror countermeasures and because of NIAID's institutional strengths in sponsoring and executing basic research, the committee considered two options for delegating a larger role to NIH for developing biowarfare countermeasures. One proposal preserved a core of DoD-based activities for budgetary and programmatic planning and management, while directing funding to research and development activities conducted by NIH and others outside DoD. The other proposal shifted all funding and programmatic responsibilities for the development of medical biodefense countermeasures from DoD to NIH, with a DoD advisory committee providing guidance to NIH on military needs and priorities The committee rejected these options in favor of recommending a DoD agency with the requisite resources for the task. Table 2-1 summarizes the advantages and disadvantages of these alternative approaches.

Minor Modifications of the Existing DoD Program

The committee considered the merits of limited changes that would preserve the basic outlines of the existing DoD framework (see Figure 2-1) while attempting to address the serious organizational and budgetary disconnect between investigators responsible for the discovery and early development of candidate countermeasures and those responsible for advanced development of these products. As described earlier in this chapter, both sets of activities are funded through the Chemical and Biological Defense Program, but they are conducted by organizations—USAMRMC and the Joint Program Executive Office for Chemical and Biological Defense—that operate in two separate Army chains of command.

Discussions at the committee's public meetings emphasized the difficulties that this organizational division creates by hindering essential collaboration between basic scientists and advanced developers. The committee also heard that current practice moves candidate products into advanced development at the start of clinical testing, a stage at which many products fail or require additional refinement, work that is most readily undertaken by the scientists involved in preparing a candidate product for clinical testing.

The change considered by the committee in this option was to shift responsibility and funding for Phase 1 clinical studies from DVC and the Joint Program Executive Office to USAMRMC. Delaying the transfer of candidate products to advanced development until a later stage of clinical testing would reduce the risk of failure of an acquisition program because only those candidates that remain viable at the end of Phase 1 testing would be transferred. This change would also allow for more complete development of associated assays and reagents crucial for testing a vaccine or drug.

Under this scenario, budgets for advanced development would also include some funding to support early collaboration by DVC with USAMRMC scientists on product testing and scale-up considerations. In addition, all DoD units with responsibility for managing aspects of the medical biodefense countermeasures program would be expected to enhance staff expertise in vaccine or drug development or regulation.

Despite having the advantage of causing the least disruption in a program that has undergone many changes over the past decade, the committee unanimously rejected this incremental change as inadequate to overcome critical underlying problems, including organizational fragmentation, lack of appropriate expertise among program leaders and managers, and insufficient visibility and priority for the program.

Consolidating Activities Within an Existing DoD Organization

Also considered was the possibility of consolidating all the activities for research and development of medical countermeasures under one of various existing organizations within DoD. This option holds the advantage of working within an already functioning administrative structure

TABLE 2-1 Advantages and Disadvantages of Alternative Approaches to the Organization and Management of Research and Development of Medical Biodefense Countermeasures to Meet DoD Needs

	New DoD Agency with Intramural and Extramural Program	DoD Agency Managing Extramural Program
Advantages	Integrated planning and management of all stages of product development DoD control of program priorities Increased visibility and priority for development of medical biodefense countermeasures Increased expertise among program leadership and managers Enhanced coordination with NIH work on bioterrorism countermeasures Expanded access to extramural researchers	Integrated planning and management for research and development of medical biodefense countermeasures Some increased visibility and priority for development of medical biodefense countermeasures Increased expertise among program leadership and managers Coordination with NIH work on bioterrorism countermeasures encouraged Expanded access to extramural researchers
Disadvantages	Disruption of establishing a new agency Potential difficulty of attracting qualified director and agency staff	Disruption of establishing a new agency Potential difficulty of attracting qualified director and agency staff Limited DoD control of program activities Loss of direct access to USAMRIID

and possibly being somewhat less disruptive than establishing a new agency from the ground up.

The committee saw little other benefit, however. An existing organization also has an existing mission and culture that would have to adapt to not only accommodate but actively support the challenging task of developing medical biodefense countermeasures. Furthermore, the committee's recommendation that management of all phases of countermeasure development, from basic science through FDA licensure, be con-

NIH/DHHS Program with DoD Advisory Committee	Consolidated Unit under Existing Organization in DoD	Minor Changes within Existing DoD System
Strong resources for managing intramural and extramural basic science Substantial NIH funding for research and development of related bioterrorism countermeasures	Integrated planning and management of all stages of product development DoD control of program priorities Does not necessitate new agency	No change in organizational structure DTRA/USAMRMC funding of initial clinical testing reduces risk of failure after candidate products transition to acquisition system DoD control of program priorities
No DoD control of medical biodefense priorities or activities Reduced visibility of medical biodefense countermeasures Likely loss of medical biodefense expertise in DoD Loss of direct access to USAMRIID Limited NIH experience with product development Limited NIH experience with military medical needs	Disruption of establishing a new unit within an existing organization Limited or no increase in program visibility Potential for incompatibilities of mission Difficulty of attracting qualified unit leader and staff	No increase in program visibility Organizational fragmentation unresolved Lack of appropriate expertise among program leadership and managers

solidated under a single director is a departure from DoD's standard practices. An existing organization that is designed to support those standard practices could impede implementation of changes the committee considers essential.

Other concerns with this approach were noted earlier in the chapter. It fails to demonstrate a true change in the priority assigned to the development of medical countermeasures or to increase the visibility of the program or its prospects of successfully competing for adequate funding.

Embedding the medical biodefense countermeasures effort within an existing organization could also make it more difficult to recruit leadership of the caliber needed for success.

Alternatives That Remove Program Components from DoD

DoD Funding and Management of an Extramural Research and Development Program

The committee considered the option of an approach that combined DoD-based program planning, budgeting, and management with extramural execution of *all* research and development tasks. This option had many of the features of the committee's primary recommendation, including creation of a DoD agency that reports to the Under Secretary of Defense for Acquisition, Technology, and Logistics. However, the research and development program would be exclusively extramural, rather than giving the agency director flexibility to carry out research through intramural or extramural programs as in the committee's recommended option. It would still be necessary for the agency director and staff to have comprehensive expertise in the management of vaccine and drug research and development (see Box 2-2) and to have the advice of external experts, but none of the research or development would by carried out by DoD personnel.

Under this scenario, the director of the DoD agency would be expected to formulate a budget to appropriately fund the program's activities. It is likely that the initial budget for this alternative would have to be at least as large as the budget using the committee's recommended approach.

Programmatic goals would be met by commissioning and funding research through NIAID and by directly supporting research and development efforts in academia and industry, using all available funding mechanisms. As part of the shift to an entirely extramural execution of research and development, responsibility and funding for the operation of the USAMRIID laboratory facility would be transferred to NIAID. The committee would, however, expect that some of the work commissioned by DoD would require use of the USAMRIID facility. Therefore, the budget for the new DoD agency would include funding to help support the facility's renovation or replacement.

DoD would rely on industry to develop candidate countermeasures through advanced development to FDA licensure and subsequent production of the products. The prime systems contractor relationship with DVC provides an established, though as yet unproven, mechanism for engaging industry services for this purpose. DoD would also be free to seek relationships with a broader range of companies.

An NIH-Based Program with a DoD Advisory Committee

The committee also discussed the option of complete termination of the DoD research and development program and transfer of its funding and programmatic responsibilities to NIH. Included in the transfer to NIH would be responsibility, funding, and staff for management and operation of USAMRIID. Guidance on DoD needs would be provided to NIH through a DoD-appointed advisory committee.

Work on biowarfare countermeasures might be assigned to NIAID, which is now the principal base for federally funded research related to medical countermeasures against bioterrorism, or to a newly created institute focused specifically on military biomedical research needs or on biowarfare and bioterrorism defense more broadly. While creation of a new institute would pose some difficulties, any overlap in research interests between this institute and NIAID would be resolved through the same types of interinstitute negotiations that already take place to resolve overlapping interests among existing NIH institutes.

A mechanism would be required to allow unique DoD needs and priorities to be represented in such an NIH-based program, for example through a DoD-appointed advisory committee composed of senior DoD officials and independent scientific experts from academia and industry.

Strengths and Weaknesses of Removing Program Components from DoD

The two alternatives that involve removing program components from DoD may have some advantages compared to the committee's primary recommendation, but they have considerable disadvantages as well. Both would capitalize on the strengths of the nation's leading agency for biomedical research, ensuring strong scientific leadership for that work and access to cutting-edge research through both intramural and extramural projects. Bringing work on biowarfare countermeasures into NIH would also promote collaboration and coordination with the closely related work on medical countermeasures against bioterrorism. The substantial funding commitment supporting the NIH activities aimed at defending against bioterrorism can be expected to allow NIH to foster a more productive environment for both intramural and extramural research than is possible with the more limited funding available within DoD. Furthermore, in the face of stiff competition for highly skilled scientists and technicians, DoD would no longer have to face the challenge of attracting and retaining the workforce needed to conduct a high-quality intramural research program. A program that is entirely NIH-based would also avoid the pitfalls of the DoD acquisition system, which is designed to oversee the development of engineering and weapons systems rather than pharmaceutical products.

However, the committee expects close coordination and collaboration between the new DoD Medical Biodefense Agency and NIAID and NIH as a whole to be the rule as it carries out its mission. Since there is considerable overlap between the biological agents of concern for both terrorism and warfare, the agency will be expected to work closely with NIH to ensure that DoD benefits from research jointly applicable to both. By maintaining the reservoir of specialized expertise that the DoD intramural research program currently represents, the Medical Biodefense Agency can benefit from flexibility to draw on its internal program or on extramural efforts as most appropriate for DoD-specific needs.

Because of crucial distinctions between biowarfare and bioterrorism and differences in the need for medical countermeasures appropriate for each, the option in which responsibility for the development of biowarfare countermeasures is given to NIAID or another entity within NIH has certain drawbacks. One concern is that it would tend to reduce the expertise in medical biodefense and the development of medical countermeasures available within DoD to guide overall biodefense planning. Furthermore, NIH has little history of focusing on military-specific needs, and among many competing national public health priorities this additional task may not be given sufficiently high priority. An NIH-based program for the development of medical countermeasures would limit DoD's influence over the research priorities for these products. The proposed advisory committee is intended to ensure that DoD's interests are known to NIH decision makers, but DoD would have no direct authority to alter NIH decisions.

The transfer of USAMRIID to NIH would be likely to bring additional scientific talent to this laboratory. Its biocontainment facilities for research on dangerous pathogens and its facilities for studies of aerosol exposures of animals are uniquely suited to biodefense research. As a part of NIH, the USAMRIID facilities could become available to support the larger national biodefense research program. However, USAMRIID is a military medical resource beyond its contribution to biowarfare countermeasure development, and its wholesale transfer from DoD would be a significant loss for the military medical community. Furthermore, despite the talent and resources USAMRIID represents, NIH and the Department of Health and Human Services may not wish to shoulder costs for the facility's urgently needed renovation.

Ensuring that biodefense vaccines and other pharmaceutical products move effectively toward FDA licensure and production is another important concern with an NIH-based program. Despite its strength in basic research, NIH has little tradition of product development (Fauci, 2003). In the area of infectious diseases, DoD, through USAMRMC, has in the past successfully collaborated with industry to gain licensure for vaccines with commercial potential (e.g., against hepatitis A, Japanese encephalitis, typhoid fever). For NIH, however, the traditional focus has been on supporting basic science and the preclinical and early clinical testing of candidate drugs and vaccines. The later stages of clinical testing and product development have generally been left to case-by-case transitions arranged between NIH or academic scientists and the private sector. This is a problematic approach for biodefense products because the limited commercial market, along with concerns about liability risks, means that the pharmaceutical industry has had insufficient incentive to seek to license and manufacture these products.

REFERENCES

AIBS (American Institute of Biological Sciences). 2001. Peer Review to the United States Army Medical Research and Materiel Command Therapeutics Research Program. Sterling, VA: AIBS.

Defense Science Board. 2002. *2001 Summer Study on Defense Science and Technology*. Washington, DC: Office of the Under Secretary of Defense for Acquisition, Technology, and Logistics.

DiMasi JA, Hansen RW, Grabowski HG. 2003. The price of innovation: new estimates of drug development costs. *Journal of Health Economics* 22(2):151–185.

DoD (Department of Defense). 2001. *Report on Biological Warfare Defense Vaccine Research and Development Programs*. Washington, DC: Department of Defense. [Online]. Available: http://www.acq.osd.mil/cp/bwdvrdp-july01.pdf [accessed February 19, 2004].

DoD. 2002. Public Law 107-107, Section 1044(e) Report to Congress: Acceleration of Research, Development, and Production of Medical Countermeasures for Defense Against Biological Warfare Agents. Washington, DC: Department of Defense, Office of the Deputy Assistant to the Secretary of Defense for Chemical and Biological Defense.

DoD. 2003. *Department of Defense Directive: The Defense Acquisition System*. Number 5000.1. Washington, DC: Department of Defense.

DVC (DynPort Vaccine Company). 2003. DVC's approach to the advanced development of medical countermeasures for biodefense. Presentation to the Institute of Medicine and National Research Council Committee on Accelerating the Research, Development, and Acquisition of Medical Countermeasures Against Biological Warfare Agents, February 20. Frederick, MD.

Fauci AS. 2003. NIAID biodefense research and collaborations with DoD. Presentation to the Institute of Medicine and National Research Council Committee on Accelerating the Research, Development, and Acquisition of Medical Countermeasures Against Biological Warfare Agents, Meeting III. Washington, DC.

GAO (General Accounting Office). 1999. *Best Practices: Better Management of Technology Development Can Improve Weapon System Outcomes*. GAO/NSIAD-99-162. Washington, DC: General Accounting Office.

Interagency Working Group on Weapons of Mass Destruction Medical Countermeasures. 2003. Terms of Reference. Washington, DC: National Science and Technology Council, Committee on Homeland and National Security.

IOM (Institute of Medicine). 2002. *Protecting Our Forces: Improving Vaccine Acquisition and Availability in the U.S. Military*. Lemon SM, Thaul S, Fisseha S, O'Maonaigh HC, eds. Washington, DC: The National Academies Press.

Pace P. 2003. Charter for the Joint Requirements Office for Chemical, Biological, Radiological and Nuclear Defense. Washington, DC: Department of Defense, Joint Chiefs of Staff.

Partnership for Public Service. 2003. *Homeland Insecurity: Building the Expertise to Defend America from Bioterrorism*. Washington, DC: Partnership for Public Service. [Online]. Available: http://www.ourpublicservice.org/publications3735/publications_show.htm?doc_id=181630 [accessed July 8, 2003].

Public Citizen. 2001. *Rx R&D Myths: The Case Against the Drug Industry's R&D "Scare Card."* Washington, DC: Public Citizen's Congress Watch. [Online]. Available: http://www.citizen.org/publications/release.cfm?ID=7065&secID=78&catID=126[accessed June 23, 2003].

Tishler CL, Bartholomae S. 2002. The recruitment of normal healthy volunteers: a review of the literature on the use of financial incentives. *Journal of Clinical Pharmacology* 42: 365–375.

Top FH Jr., Dingerdissen JJ, Habig WH, Quinnan GV Jr., Wells RL. 2000. DoD Acquisition of Vaccine Production: Report to the Deputy Secretary of Defense by the Independent Panel of Experts. In DoD, 2001. *Report on Biological Warfare Defense Vaccine Research and Development Programs*. Washington, DC: Department of Defense. [Online]. Available: http://www.acq.osd.mil/cp/bwdvrdp-july01.pdf [accessed February 19, 2004].

U.S. Congress, Office of Technology Assessment (OTA). 1993. Pharmaceutical R&D: Costs, Risks and Rewards. OTA-H-522. Washington, DC: U.S. Government Printing Office. [Online]. Available: http://www.wws.princeton.edu/~ota/ns20/year_f.html [accessed July 8, 2003].

3

Challenges in the Research and Development of Medical Countermeasures Against Biological Warfare Agents

The policy and organizational changes that the committee has recommended for the Department of Defense (DoD) are a portion of what needs to be done to permit more effective progress toward the development and licensure of medical biodefense countermeasures. But DoD, along with others working to develop medical countermeasures, has to confront other important challenges as well. Academia and private sector firms are essential partners in government efforts to discover, develop, and manufacture medical countermeasures. However, various factors, including concerns about potential liability risks, have deterred the pharmaceutical industry from becoming involved in producing these specialized products that have limited commercial markets.

The regulatory process and the role of the Food and Drug Administration (FDA) also require attention. Determining the efficacy of biodefense products poses special challenges because ethical constraints prevent efficacy testing in humans. The adoption in 2002 of the "Animal Efficacy Rule," the regulatory mechanism that permits efficacy data to be obtained from tests using animals (FDA, 2002b), opens a pathway to licensure. However, extensive research and testing will be needed to guide the application of this new regulatory tool. Furthermore, expediting the testing and review of high-priority biodefense products places heavy demands on FDA resources.

In addition, accelerating the development and testing of countermeasures will require ensuring the availability of adequate supplies of nonhuman primates and other laboratory animals, specialized laboratory facilities with appropriate biosafety features, and facilities in which test lots of candidate countermeasures can be produced in compliance with FDA standards for current Good Manufacturing Practice (cGMP). The nation also faces a limited supply of scientific and technical personnel with the expertise needed to carry out the work at many stages in the development of medical countermeasures (Partnership for Public Service, 2003).

In this chapter, the committee reviews these challenges and recommends steps that DoD, acting principally through the proposed Medical Biodefense Agency, should take on its own or in collaboration with others.

ENGAGING ACADEMIA AND INDUSTRY

Under the new DoD organization recommended by the committee, partnerships with the academic community and with biotechnology and pharmaceutical companies will be crucial to the success of the medical countermeasure development program. Early research and discovery leading to new candidates for vaccine and drug countermeasures should involve both intramural and extramural work, with the mix determined on the basis of assessments by the DoD program's leadership of the most effective way to achieve programmatic goals.

With no government-owned facilities for full-scale vaccine or drug manufacturing, collaboration with experienced commercial partners is essential to move candidate countermeasures through the final stages of product development and licensure and into production. In addition, access to manufacturing expertise during the early stages of product development can help avoid wasted effort and inappropriate commitment of resources.

At present, DoD's Joint Vaccine Acquisition Program has a contract with the DynPort Vaccine Company LLC (DVC) to develop and license biological defense vaccines for the U.S. armed forces. DVC does not have vaccine production facilities of its own but instead contracts with companies possessing the necessary facilities and expertise.

Potential Obstacles to DoD Partnerships with Academia and Industry

DoD's need to collaborate with academia and industry in the development of medical countermeasures is clear. However, the discussions at the committee's public meetings, testimony to congressional committees, and comments in other contexts have highlighted factors that discourage collaboration with DoD as well as other government agencies that are exploring ways to encourage industry participation in the development and production of medical countermeasures to be used against bioterrorist threats (e.g., Aventis Pasteur, 2002; Defense Science Board, 2000; Read, 2003; Top et al., 2000).

Factors of concern to both academia and industry include the following:

- Complex and cumbersome contracting requirements
- Potential instability of government funding because of the annual appropriations process
- Exposure to financial risks for product liability claims
- Restrictions on work with pathogens designated as select agents[1]
- Lack of appreciation within DoD of the true costs of vaccine and drug development

Factors primarily influencing larger pharmaceutical firms include the following:

- Short-term opportunity costs in diverting limited manufacturing capacity and skilled personnel from other projects
- Limited potential to obtain revenue or large profit margins from government contracting
- A small or uncertain market beyond government purchases
- Antitrust restrictions that limit collaboration among firms

In light of the mixture of challenges that exist, several approaches or potential solutions have been proposed. The committee makes recommendations regarding some of these options below.

Full Use of Existing Grant and Contract Mechanisms

The committee expects the proposed Medical Biodefense Agency to make full use of all means at its disposal to ensure that the most suitable organization conducts the work needed to achieve program goals. Examples of the types of agreements that might be employed are listed in Box 3-1. The current Medical Biological Defense Research Program,

[1]As mandated by the Public Health Security and Bioterrorism Preparedness Response Act of 2002 (P.L. 107-188), the Centers for Disease Control and Prevention (CDC) must maintain a list of "select agents," that is, biological agents and toxins considered to pose a public health or agricultural threat. As directed by P.L. 107-188, CDC (2002) has established requirements regarding the possession and use of select agents. These requirements concern registration, security risk assessments, safety plans, security plans, emergency response plans, training, transfers, record keeping, inspections, and notifications.

BOX 3-1
Mechanisms for Engaging Academic Institutions, Industry, and Others in the Private Sector in Federally Funded Research and Development

- Grants
- Contracts
- Cooperative Agreements
- Small Business Innovation Research (SBIR) grants of up to $850,000 available to small, for-profit businesses for early stage research and development
- Small Business Technology Transfer (STTR) grants of up to $600,000 available to fund cooperative efforts between small businesses and U.S. research institutions for early stage research and development
- Cooperative Research and Development Agreements (CRADAs) to conduct research and development in a technical area of mutual interest to federal laboratories and a nonfederal party (industry, university, not-for-profit organization, or state or local government)
- Dual-Use Science and Technology (DUS&T) development projects under which dual-use technology research or development is carried out sharing the costs between DoD and nongovernmental entities
- Other Transactions: mechanisms other than a contract, grant, or cooperative agreement used with commercial firms that do not normally contract with DoD. They generally do not require compliance with federal laws and regulations that apply to procurement contracts, grants, and/or cooperative agreements.

SOURCES: http://www.acq.osd.mil/sadbu/sbir/overview/index.htm.; Dual Use Science & Technology Process: Why Should Your Program Be Involved? May 2002. http://www.dtic.mil/dust/dust_process_02.pdf; http://www.dtic.mil/dust/; http://www.acq.osd.mil/cp/winegar_8-8-02_missouristconf.pdf; Other Transaction Authority (OTA) for Prototype Projects. http://www.acq.osd.mil/dp/dsps/ot/dspsot.htm; all accessed 8.13.03.

through which basic research and early product testing are supported, makes use of many of these mechanisms. Roughly 40 percent of the planned budget allocation for fiscal year (FY) 2003 was for support of extramural efforts, including a portion of the core program and all of the projects requested by Congress or transferred from the Defense Advanced Research Projects Agency (DARPA) (Henchal, 2003; Linden, 2003).

The committee emphasizes in particular the usefulness of "other transaction" authority as a way to gain greater flexibility in the agreement process. This mechanism is specifically intended to facilitate collaborations with commercial firms that are not traditional DoD contractors, and it has been used successfully for this purpose by DARPA (Tether, 2002). "Other transactions" include any kind of transaction other than a contract, grant, or cooperative agreement. Their use is authorized for DoD research projects by 10 U.S.C. 2371 and for prototype projects relevant to weapons or weapons systems by section 845 of the National Defense Authorization Act for Fiscal Year 1994 (P.L. 103-160). The National Defense Authorization Act for Fiscal Year 2004 (P.L. 108-136), which was signed into law in late 2003 as this report was being completed, explicitly extends the authority for "other transactions for prototype projects" related to defense against nuclear, biological, chemical, or radiological attack. Agreements under this authority are usually exempt from the federal acquisition laws and regulations that apply to standard contracts, grants, and cooperative agreements (see DoD, no date a; 2001; GAO, 2002).

5. The Medical Biodefense Agency should fully utilize "other transactions" authority as a means of encouraging academia and private sector firms to participate in the research and development of medical biodefense countermeasures to meet DoD needs.

Stability of funding for those awarded contracts and grants is another important consideration for potential partners. The committee urges measures to enable the Medical Biodefense Agency to make long-term commitments of funds to its awardees in a manner that would expedite program execution, but would do so in an economically efficient way. The authority to sign multiyear contracts without full funding of termination liabilities would allow DoD to contract for the full scope of a project lasting more than one year, but to budget for the project over several years as fiscal obligations become due.

6. Congress should authorize the Medical Biodefense Agency to sign multiyear contracts without a requirement for full, up-front funding of any termination liabilities.

New Tools Available to DoD

DoD is not alone in its need to form partnerships with potentially reluctant private companies to obtain medical countermeasures for biodefense. Plans to acquire countermeasures for a national stockpile for defense against bioterrorism also depend on commercial firms' being willing to complete development of and produce the vaccines and drugs to be included. For much of 2003, it was expected that Congress would pass legislation to implement Project Bioshield, a Bush administration proposal to address some of the potential obstacles to participation by commercial firms and to help speed certain research management and product procurement processes.

The Project BioShield proposal included three principal features: (1) relaxing acquisition procedures for procuring property or services in support of research and development for biomedical countermeasures and expediting the peer review process for Department of Health and Human Services (DHHS) grants, contracts, and cooperative agreements; (2) allowing DHHS to contract to purchase countermeasures that are still undergoing clinical testing; and (3) providing for emergency use of countermeasures not yet approved by FDA (Gottron, 2003).

As of December 2003, legislation to authorize Project BioShield had passed in the House (H.R. 2122) but was pending in the Senate (S. 15 and S. 1504) (Gottron, 2003). However, other legislation already approved by Congress has enacted some elements of the BioShield proposal. The appropriations bill for the Department of Homeland Security (DHS)[2] makes $5,593 million available until September 30, 2013, for securing biodefense countermeasures. Of this amount, up to $3,418 million may be obligated during FY 2004 through 2008 and up to $890 million may be obligated in FY 2004.

In addition, and of particular importance for DoD, are the BioShield-like provisions of the National Defense Authorization Act for Fiscal Year 2004 (P.L. 108-136). The extension of "other transactions" authority, which is not limited to DoD, has already been noted. The legislation also includes a general provision authorizing increased financial thresholds for the use of simplified acquisition procedures "to facilitate the defense against or recovery from nuclear, biological, chemical, or radiological attack against the United States" (Sec. 1443).

Other provisions of P.L. 108-136 apply specifically to DoD. Section 1601 includes provisions authorizing the Secretary of Defense to use streamlined acquisition procedures, "when appropriate," in procuring property or services for use in performing, administering, or supporting research and development of biomedical countermeasures. (Provisions related to laboratory construction and personnel are noted later in this chapter.) Section 1602 directs the Secretary of Defense to identify and procure for a DoD stockpile the biomedical countermeasures needed to pro-

[2]Making appropriations for the Department of Homeland Security for the fiscal year ending September 30, 2004, and for other purposes, P.L. 108-90 (2003).

tect members of the armed forces against leading biological, nuclear, chemical, and radiological threats. Procurement is limited to "qualified countermeasures" that are (1) already approved by FDA or judged by the Secretary of Health and Human Services as likely to qualify for such approval and (2) considered feasible to produce and deliver in the quantities needed by DoD within 5 years. The Secretary of Defense is authorized to use an interagency agreement with DHS and DHHS to obtain countermeasures for DoD from the Strategic National Stockpile and to transfer funds to the other agencies to cover the cost of replenishing the stockpile. The legislation establishes no formal requirement that manufacturers of countermeasures obtained for the DoD stockpile ensure that those products ultimately receive FDA approval. Section 1603 is covered later in the chapter in the discussion of emergency use of medical countermeasures by DoD.

It is difficult to anticipate whether the added funding and procurement authorities enacted so far—or the BioShield proposals, if passed and signed into law—will provide sufficient incentive for industry to enter into development agreements with DoD or DHHS. However, the relaxation of acquisition procedures, expedited peer review, and market guarantees appear to be constructive steps toward addressing some of the concerns of the private sector.

Some argue, however, that additional incentives will be needed (e.g., Barbaro, 2003; Ludlam, 2003; Read, 2003; Ryan, 2003). For example, neither the DoD authorization legislation nor the BioShield bills address industry concerns regarding the need for indemnification against product liability claims. The proposed Biological, Chemical, and Radiological Weapons Countermeasures Research Act (S. 666), introduced in March 2003, is one attempt to try to address some of these other concerns, including tax and intellectual property rights considerations, liability concerns, limited antitrust exemptions, and incentives to increase research and manufacturing capacity. As of December 2003, neither the Senate nor the House had acted on this bill.

Liability Considerations

Most medical biodefense countermeasures are likely to receive FDA approval under the Animal Efficacy Rule and, thus, without direct evidence of efficacy in humans. Moreover, unlike products developed for a commercial market, most of the biodefense vaccines and some of the therapeutic products will be purchased only by government agencies and used only at the direction of those agencies. For major pharmaceutical firms, the possibility of substantial financial loss through product liability claims is a significant deterrent to their willingness to apply their expertise and resources to the development of new countermeasures. Smaller companies may initially be willing to accept greater financial risk, but they also have fewer resources to sustain their efforts to develop new products. Concerns about liability may extend as well to university researchers and other not-for-profit research organizations.

Having heard from biotechnology and pharmaceutical industry representatives and from a consumer advocate, the committee is persuaded that it is important for the government to address industry concerns about product liability risks as part of efforts to accelerate the development of medical biodefense countermeasures.

Among the possible alternatives is the approach adopted under the Homeland Security Act of 2002 for the use of the smallpox vaccine.[3] When the vaccine is administered in response to an appropriate declaration by the Secretary of Health and Human Services, the vaccine manufacturer, as well as health care organizations and individuals involved in administering the vaccine, are deemed to be employees of the Public Health Service for purposes of any liability claims. Suits for personal injuries allegedly caused by the vaccine must be brought against the federal government, which may seek recovery against manufacturers and other covered persons only for gross negligence, illegal conduct, willful misconduct, or violations of government contract provisions. This is similar to the legislation enacted in 1976 (P.L. 94-380) in response to liability and insurance concerns raised in connection with the swine flu program.

The Homeland Security Act also included the "Support of Anti-terrorism by Fostering Effective Technologies Act of 2002," referred to as the SAFETY Act. The SAFETY Act includes a provision limiting the tort liability to sellers of designated antiterrorism technologies and providing for the use of the "government contractor defense,"[4] after approval from the Secretary of Homeland Security (Department of Homeland Security, 2003). However, the limits under the SAFETY Act do not apply to harm caused when no act of terrorism has occurred, so this provision may not cover vaccines that might be used when an attack is only suspected or threatened (Gottron, 2003) or vaccines administered to U.S. troops to pro-

---

[3]Liability protections related to the smallpox vaccine were included in the Homeland Security Act of 2002 (P.L. 107-296), passed in November 2002, and in amendments enacted in the Smallpox Emergency Personnel Protection Act of 2003 (P.L. 108-20), passed in April 2003.

[4]The government contractor defense is a judicially created doctrine barring claims against government contractors who meet certain requirements. The SAFETY Act provides that this defense is available upon approval by the Secretary of Homeland Security to those who sell to state and local governments and the private sector as well as to the federal government.

tect against potential battlefield (rather than bioterrorist) exposure to biological agents.

Another model is the National Vaccine Injury Compensation Program (P.L. 99-660), which is a federal "no-fault" system to provide compensation for injuries related to specific childhood vaccines. This approach requires sufficient prior knowledge of the products and their associated adverse effects so that a list of covered injuries can be established. As a result, this model is not practicable for as-yet undeveloped vaccines and drugs for which adverse event profiles have not been determined.

Another approach is contractual indemnification, which obligates the federal government to pay a contractor's costs incurred as a result of litigation. Through its prime system contract with DVC, DoD already makes indemnification available, on request, to DVC's subcontractors as an extension of the indemnification provided to DVC. However, because companies can recover litigation costs only after they are incurred, this approach can allow companies to fail before they are able to receive payment from the government.

The committee favors an approach similar in concept to the Homeland Security Act model for the smallpox vaccine. The government would be the sole defendant in any suit alleging injury in connection with a designated biodefense countermeasure, but would retain the right to seek indemnification from a manufacturer or other covered person for specified actions, such as gross negligence, willful misconduct, or criminal acts. Experience with the swine flu program in 1976 suggests that this will provide sufficient assurance for large pharmaceutical companies and others to participate in the development and manufacture of the needed products, while retaining incentives for all participants to exercise appropriate care in the research, development, and manufacturing process. This approach will require some form of legislation. Until such legislation can be enacted, the committee recommends that DoD and DHHS make maximum permitted use of existing legislative authority to enter into indemnification agreements with persons who contract for research, development, and manufacture of biodefense countermeasures.

> **7. DoD and DHHS should make maximum permissible use of statutory indemnification authority under existing legislation to encourage entities in the private sector, including universities and other research institutions and companies, to enter into agreements to develop and manufacture medical countermeasures against biowarfare agents. As soon as possible, legislation should be enacted creating a system comparable to that for the smallpox vaccine under the Homeland Security Act, under which suits for personal injuries allegedly caused by biowarfare countermeasures may be**

brought only against the federal government, which would retain the right to recover damages resulting from such suits from manufacturers or other covered persons if their misconduct (gross negligence, illegal acts, willful misconduct, or violation of government contract obligations) was shown to be the cause of the injuries.

SUPPORTING THE REGULATORY PROCESS

FDA has statutory responsibility for ensuring the safety and efficacy of drugs and biologics approved for human use,[5] which gives the agency a crucial role in the development and licensure of medical biodefense products of interest to DoD. As a result, efforts to accelerate the development of medical countermeasures for DoD have to take into account factors related to FDA. The committee focused on three issues that require further attention: application of the new Animal Efficacy Rule, ensuring that FDA has the resources necessary to expedite the review of product license applications for medical countermeasures, and planning for limited or emergency use of medical countermeasures.

Using the Animal Efficacy Rule

The Animal Efficacy Rule, finalized by FDA in 2002, enables FDA to approve new vaccines and drugs when it is not ethical or feasible to conduct human efficacy studies, a situation that applies to most biodefense products. Under this new rule, evidence of efficacy can be based on data derived from studies in more than one animal species "expected to react with a response predictive for humans" or in a single sufficiently well-characterized animal model (FDA, 2002b). In February 2003, FDA announced the first product approval under the Animal Efficacy Rule: pyridostigmine bromide for pretreatment prophylaxis against exposure to the nerve agent soman (FDA, 2003a).

A substantial amount of work must be done to validate animal models as surrogates for humans in efficacy studies of biowarfare countermeasures. It will be necessary to establish the biologic plausibility of the equivalence of animal and human responses to the disease agents and to the countermeasures in question. DoD-sponsored investigators have helped develop some of the few existing animal models of human disease caused by biological warfare agents or of protection against those dis-

[5]Federal Food Drug and Cosmetic Act Section 505i; The Public Health Service Act, Regulation of Biological Products, Section 351 Subpart 1.

eases (e.g., Geisbert et al., 2002; Ivins et al., 1996, 1998; Jahrling, 2002; Pitt et al., 1996, 2001; Zaucha et al., 2001), but the National Institute of Allergy and Infectious Diseases (NIAID, 2002) has identified a need for additional work to develop animal models for almost all of the biological agents considered to pose the most serious threat for bioterrorism (referred to as "Category A" agents).[6]

Although the committee believes that NIAID should be the primary sponsor of such work, the Medical Biodefense Agency should ensure that information on animal models developed by DoD investigators is available to researchers and the product development community, including FDA. The Medical Biodefense Agency should also have adequate funding to help support development of the animal models that will be necessary to establish the efficacy of biodefense products intended to meet unique DoD needs.

8. The Medical Biodefense Agency and the National Institutes of Health (NIH) should cooperate in making information on animal models relevant for the development of medical biowarfare countermeasures available to qualified investigators. The DoD agency should work with NIH and engage FDA to develop additional animal models that will be useful for specific agents or products of particular concern to DoD. The Medical Biodefense Agency should receive funding specifically for this task.

The Animal Efficacy Rule is new, and there is still uncertainty about its interpretation and application. Although FDA will be the final judge of the data needed to provide evidence of efficacy, advances are required in the scientific field as a whole, accompanied by scientific discussion and consensus building regarding the best approaches. FDA should receive funding to support this additional work, which will help establish an essential scientific foundation for use of the Animal Efficacy Rule.

9. FDA should work with the scientific community to enrich the science base that the agency will have to draw on in order to apply the Animal Efficacy Rule. FDA should receive sufficient funding to support both intramural and extramural work on these issues.

---

[6]The CDC has established three categories of biological agents that are considered to be bioterrorism threats (http://www.bt.cdc.gov/agent/agentlist-category.asp). Category A agents are those considered to pose the most serious risk to national security; these agents are listed in Appendix A.

The Pace of FDA Review

As noted in Box 1-2, a sponsor's Investigational New Drug (IND) application must be acceptable to FDA before Phase 1 clinical trials begin, as must plans for subsequent Phase 2 and Phase 3 testing. Once clinical testing is complete and manufacturing processes are established, a sponsor must file an application for marketing approval. The completeness and quality of the sponsor's submissions to FDA are crucial elements in the speed with which FDA accepts plans to move forward with testing or completes the review of a product license application. In addition, FDA resources and options for rapid review play a role. Two mechanisms for expedited review of any qualifying product license application are currently available.

Fast-track status (sometimes called subpart H or accelerated approval) is available for preventive products or drugs that provide meaningful therapeutic benefits over existing treatments for serious or life-threatening illnesses (21 U.S.C. 356; FDA, 1998). It permits marketing approval on the basis of clinical trials using surrogate endpoints reasonably likely to predict clinical benefit or on the basis of an effect on a clinical endpoint other than survival or irreversible morbidity, often in conjunction with requirements for postmarketing studies.

The Prescription Drug User Fee Act (PDUFA), passed in 1992 and renewed in 1997 and 2002, provides another mechanism for expedited review (21 U.S.C. 379). It authorizes FDA to collect fees from sponsors of New Drug Applications (NDAs) and Biological License Applications (BLAs) for innovative drugs and biological products, as well as annual fees for products and manufacturing facilities. Income from these fees is largely earmarked to provide resources for the FDA review process. Under the most recent PDUFA agreement, FDA is committed to act on 90 percent of standard NDAs and BLAs within 10 months and 90 percent of "priority" applications within 6 months. FDA's action does not, however, necessarily mean approval within this period of time.

To speed action on medical countermeasures specifically, FDA has adopted practices that are more proactive than those followed for most other types of products. For example, even before the bioterrorist use of anthrax in the U.S. mail in 2001, FDA had begun reviewing data for existing products to facilitate the licensure of countermeasures against biological, chemical, and nuclear agents. In 1999, FDA invited the manufacturer of ciprofloxacin to submit an application to modify the product labeling to include an indication for use in cases of inhalational exposure to *Bacillus anthracis* (FDA, 2000). In November 2001, in response to concerns about possible shortages of ciprofloxacin, FDA drew on its review of the data on doxycycline and penicillin G procaine products to publish a

Federal Register notice clarifying that these products were also approved for use following inhalational exposure to *B. anthracis* and providing dosing regimens (FDA, 2001). The same notice also encouraged manufacturers to submit applications to change the labeling of their products to reflect the new dosing information. In March 2002, FDA published a draft guidance document on the development of drugs to treat exposure to inhalational anthrax (FDA, 2002a), and in April 2002, FDA joined with DoD to sponsor a workshop to discuss strategies for testing investigational anthrax vaccines, including the identification of surrogate markers.

In addition, FDA has been increasingly proactive and helpful in its product-specific discussions with sponsors as they bring candidate vaccines and drugs forward for clinical testing. FDA's accessibility is especially valuable to newer companies that are seeking approval of products for the first time and are often less familiar than larger, well-established pharmaceutical firms with FDA procedures and requirements. In discussions with the committee during its information-gathering meetings, both DoD and industry representatives noted the agency's proactive practices related to biodefense countermeasures.[7]

FDA's expanded efforts require additional staff time, which translates into the need for more staff or the diversion of staff from other tasks. The committee understands that FDA is giving biowarfare or bioterrorism countermeasures de facto priority status. These extraordinary efforts have been made in response to urgent national need, but they are not sustainable without additional resources, particularly in the form of trained personnel. Although FDA has already received some additional funding and personnel, the committee believes it crucial that Congress ensure that funding to support the additional work that FDA is doing in response to the threats of biowarfare and bioterrorism continues to be sufficient to allow the agency to sustain its efforts.

10. Congress should ensure that adequate funding is provided to support the additional work that FDA is carrying out in response to threats from bioterrorism and biowarfare.

A Need for Contingency Planning

Ideally, the biowarfare countermeasures currently being planned or developed will have been licensed by FDA before they are needed to pro-

[7]FDA issues were a focus of discussion at the committee's March 2003 meeting. See Appendix B for agendas for the committee's information-gathering meetings.

tect troops about to be deployed or those already on the battlefield, but plans should be made concerning two issues related to product approvals. One issue is the possibility of speeding DoD access to a countermeasure by approving a product for limited use (e.g., by healthy adults) or limited distribution, which FDA already has the authority to do. Another concern is the possibility that an urgent need will arise for one or more countermeasures before they have been licensed for use by the FDA. In such a circumstance, careful assessment of the risk of the threat and the consequences of having no medical countermeasure would have to be weighed against the remaining unknowns regarding the safety and efficacy of the product in question.

Special Use Licensure

In decisions regarding licensure of medical countermeasures and other products, FDA has to take into account the health status of the population for whom the product is intended. In evaluating treatments for diseases, FDA must consider whether the treatment is likely to improve the health of the patient and whether the expected benefits of the treatment outweigh the risk of adverse health effects. Vaccines are held to a particularly high standard of safety because they are usually given to healthy people to protect them against a disease to which they may never be exposed. In contrast, therapeutic agents are given when a disease is already known to be present, or is at least suspected.

As part of the labeling for licensed products, FDA requires a package insert that provides relevant clinical information, including the population for whom the drug or biologic is appropriate. The agency can specify, for example, that a countermeasure is indicated only for persons in a particular age group, with a particular health status, or in an occupation that entails a high risk of exposure to a particular hazard. For example, the new nasally administered influenza vaccine was approved for use by healthy persons 5 to 49 years old (FDA, 2003b). Other indications for limited use can also be specified. With pyridostigmine bromide, for example, the recently approved labeling indicates that its use as a pretreatment against nerve agent exposure is "for military combat medical use only" (FDA, 2003c). Postlicensure studies can provide additional safety data to factor into consideration of future labeling changes.

Through this means, countermeasures that have been adequately tested in young and healthy populations might be licensed with labeling specifying use only in such populations and only when risk of exposure is sufficiently high to warrant accepting the potential risks. Such licensure would make products available for DoD use before the additional testing that would be needed to license products for the general population, including children, the elderly, and those with compromised immune systems.

Like pyridostigmine bromide, some biodefense products might also be approved specifically for military use. One concern with this approach is that such approvals could be thought to suggest that military personnel were being given products that were not good enough for use in the general population. The questions raised about possible adverse health effects from vaccines and drugs given to troops during the 1990–1991 Gulf War illustrate the nature of this concern. The committee agrees that the preferred approach, even for countermeasures that are likely to be used only by DoD, would be to seek approval for use in as broad a population as possible.

Emergency Use of Unlicensed Countermeasures

The committee fully supports the DoD policy that use of FDA-approved products is the preferred means of providing force health protection (DoD, 2000). However, given the number of biodefense products under development, it is likely that some will not have been licensed at the time they are needed and will have to be used under IND provisions. The IND provisions are useful for protecting research subjects, but are not ideal for administering a vaccine or drug to a large military contingent to provide force health protection, particularly because of the requirement to obtain and document informed consent.

The issue is especially challenging because, at the time of the Gulf War of 1990–1991, DoD requested and was ultimately granted a waiver of the informed consent requirements for the use of two investigational products, pyridostigmine bromide and botulinum toxoid vaccine, on the grounds that informed consent was not feasible for military exigencies (FDA, 1990; Rettig, 1999). With the subsequent incidence of unexplained illness among military personnel who served in the Gulf, which some thought should be attributed to the medications they received, this waiver proved highly controversial (for a more complete discussion, see Rettig, 1999).

Congress has since passed a law specifying that only the President may grant a waiver of informed consent to use an investigational product for force health protection in connection with service members' participation in military operations (10 U.S.C. 1107). In the absence of a Presidential waiver of informed consent, current DoD policy specifies that when no FDA-approved drug or biological product is available at the time of the need for a countermeasure against a particular threat, DoD components may use an investigational product only with the approval of the Secretary of Defense and only in compliance with requirements for informed consent by personnel receiving the product. The request to the Secretary must document a confirmed, high threat for which the use of an investigational drug or biologic product is needed; consideration of the risks and benefits of use of the investigational product; and compliance with the DoD directive spelling out record-keeping and other associated requirements (DoD, 2000). A treatment protocol must be developed for the use of the investigational product and submitted to FDA for review after approval by a duly constituted institutional review board.

With the passage of the National Defense Authorization Act for Fiscal Year 2004 (P.L. 108-136), DoD has been given an additional mechanism for gaining approval for emergency use of medical countermeasures that have not been approved by FDA or have not been approved for a specific use. With a determination by the Secretary of Defense that there is a military emergency, or significant potential for a military emergency, involving a heightened risk to U.S. military forces of attack with a specified biological, chemical, radiological, or nuclear agent, the Secretary of Health and Human Services may authorize distribution and use of such countermeasures.

To exercise this authority, the Secretary of Health and Human Services must conclude the following: (1) the agent for which the countermeasure is designed can cause a serious or life-threatening disease; (2) the product may reasonably be believed to be effective in detecting, diagnosing, treating, or preventing the disease; (3) the known and potential benefits of the product outweigh its known and potential risks; (4) no adequate alternative to the product is approved and available; and (5) any other criteria prescribed in federal regulations are met.

The statutory requirements governing the administration to military personnel of investigational biologics or drugs (10 U.S.C. 1107) would not apply to use of products under such an emergency declaration. However, military personnel receiving these countermeasures are still to be informed that the Secretary of Health and Human Services has authorized emergency use of the product. They are also to be informed of potential benefits and risks of the product and the extent to which those benefits and risks are known, of any option to refuse the product, and of available alternatives.

If the President determines, in writing, that complying with the requirement to provide information on the option to accept or refuse administration of a product is not feasible, is contrary to the best interests of the service members affected, or is not in the interests of national security, the President may waive it. However, this information must be provided to service members (or their families) not more than 30 days later. In addition, information concerning administration of the product is to be recorded in the service member's medical record.

The legislative proposals for Project BioShield include similar provisions for authorizing emergency use by military personnel of medical countermeasures that are not yet approved by FDA, as well as provisions for authorizing emergency use of such countermeasures in the civilian population. The status of these bills (S. 15, S. 1504, and H.R. 2122) was uncertain at the time the committee completed this report. If such legislation is enacted, P.L. 108-136 specifies that its emergency use provisions are to be replaced by those in the BioShield legislation upon notification of Congress by the President that those provisions provide effective emergency use authority for DoD.

The committee considers it appropriate, and necessary, for DoD to have mechanisms such as these for using unlicensed countermeasures without the constraints of informed consent in those instances when a biological threat is considered sufficiently great, contingent upon approval from the President or the concurrence of the Secretary of Health and Human Services. Ensuring that DoD can respond in an effective and timely manner to such emergencies will require ongoing planning and coordination among various components within the department, including the Medical Biodefense Agency with its proposed responsibility for managing the development of medical countermeasures and the Office of the Assistant Secretary of Defense for Health Affairs, which has policy responsibilities for immunization and use of IND products (DoD, 1993, 2000).

OVERCOMING CURRENT AND POTENTIAL RESOURCE BOTTLENECKS

Research and development of medical countermeasures against biowarfare agents require an array of resources, some of which are beyond those needed for routine biomedical research and pharmaceutical product development. These resource needs span the process from discovery to licensure. Of particular importance for the development of medical biodefense countermeasures is the availability of nonhuman primates to serve as test subjects, specialized laboratory and animal testing facilities, and facilities for producing candidate countermeasures in compliance with cGMP standards. These are resources that often cannot simply be purchased.

Because of recent substantial investments by NIAID in the early stages of product research (discovery), resource constraints may be felt most acutely when increased numbers of candidate countermeasures reach the stages of advanced research and development. Steps should be taken to help ensure that each of these critical components is available as required.

A Need for Coordination

Although at one time DoD was the only organization carrying out research and development activities aimed at drugs and vaccines to protect against biological warfare agents (many of which are also bioterror agents), several federal agencies and academic institutions now have ongoing or proposed roles. Large increases in funding, primarily through NIH, are fueling the startup of many academic research programs across the country. Federal agency interest in various aspects of research to protect against bioterror agents has also expanded to DHS, the Department of Agriculture, and the Department of Justice. Private companies are demonstrating increased involvement in this research and development area, as well. For example, using its own funding, Human Genome Sciences has developed a candidate monoclonal antibody to protect against toxin produced by *B. anthracis* (Albert, 2003; Gillis, 2003).

With so many likely participants in biodefense research and the efforts to develop medical countermeasures against biological agents, the demand for crucial resources is growing, and some shortages are anticipated. As a result, it is imperative that the existing resources be used prudently and that their use be coordinated to the maximum extent possible. Systematic planning and assessment should guide action to meet future needs. Within the past year, a new Interagency Working Group on Weapons of Mass Destruction Medical Countermeasures has begun to provide a forum for interagency coordination in several different areas, including medical countermeasures against biological warfare and bioterrorism agents. DoD, DHHS, and DHS are participants, as well as several other departments. In addition, an interagency agreement between NIAID and the U.S. Army Medical Research Institute of Infectious Diseases (USAMRIID) provides for collaborative efforts and use of facilities and personnel in joint activities, and for the construction of additional biocontainment space for nonhuman primates in the USAMRIID facilities at Fort Detrick in Maryland (NIAID and USAMRIID, 2002). Coordination of the separate efforts of federal agencies is fraught with challenges but should be pursued to prevent both unwarranted redundancy and critical shortages. The committee further emphasizes the need for coordination in the particular resource areas discussed below.

Resources for Product Research and Development

Availability of Nonhuman Primates

Studies using nonhuman primates are likely to be important in meeting the requirements of the new Animal Efficacy Rule because of the animals' similarities to human beings. A variety of nonhuman primates have been used in biomedical research, but the Indian-origin rhesus macaque is frequently favored because information on the biology of the species has accumulated from more than 50 years of use (Hearn, 2003). One of the resource concerns for the development, testing, and evaluation of medical biodefense countermeasures is an ongoing and critical shortage of Indian-origin rhesus macaques. This shortage is a product of a variety of factors, including heavy demands from HIV investigations and other biomedical research, difficulties in importing and transporting the animals, and limited holding space and breeding capacity in the United States (Black, 2003; DeMarcus, 2003; NRC, 2003; Personal communication, B. Weigler, Fred Hutchison Cancer Research Center, May 28, 2003). The infusion of funding through NIH for the development of bioagent countermeasures may intensify the shortage (Shortage of monkeys, 2003).

However, several other species of nonhuman primates are likely to be useful in this work. Alternatives to the Indian-origin rhesus macaque, such as the Chinese-origin rhesus macaque and the cynomolgus macaque, should be studied and used where possible. A large body of knowledge already exists for the cynomolgus macaque (Bennett et al., 1995; NRC, 2003; Research Resources Information Center, no date), which was used in recent studies on a candidate vaccine against Ebola virus (Sullivan et al., 2003). The committee endorses the continuation of work like that supported by DARPA to validate other nonhuman primate models of disease and countermeasure activity in nonhuman primates other than Indian-origin rhesus macaques (Carney, 2003).

For all of these models, efforts will be needed on several fronts to make effective use of available animals and, over the longer term, to appropriately expand their supply. It is important that the Medical Biodefense Agency, NIH, and others involved in biodefense research participate in and provide financial support for coordinated efforts to increase domestic cage and breeding capacity as part of efforts to expand the supply of nonhuman primates. With the growth of medical biodefense research, it is imperative for the Medical Biodefense Agency, NIH, and others involved in this work to assess their current and future needs for nonhuman primates and to coordinate their use of the available animals.

11. The Medical Biodefense Agency should participate in a national effort to support the maintenance and expansion of nonhuman primate research resources, which will be critical to the success of efforts to develop medical biodefense countermeasures. The Medical Biodefense Agency should be provided with sufficient funding for these activities.

Facilities for Research and Product Development and Testing

Specialized facilities for research, development, testing, and evaluation are part of the essential infrastructure for the development of medical biodefense countermeasures. Work with most of the Category A agents requires facilities with high-level biosafety ratings (biosafety level [BSL] 3 or 4), putting the availability of these facilities on the critical path for countermeasure development. Furthermore, the military threat posed by most biological agents is generally considered to be from aerosol exposure. As a result, efficacy testing for candidate countermeasures for DoD requires facilities in which test animals can be subjected to aerosol challenge under high-level biocontainment conditions.

At the present time, only a few BSL-3 and BSL-4 facilities are operational, including those at USAMRIID. USAMRIID and Battelle Memorial Institute have the nation's primary facilities for aerosol challenge studies of nonhuman primates. As a result of increases in the number of candidate countermeasures expected to be moving through the development pipeline, there is a need for additional biocontainment space for laboratory research, for animal experiments involving aerosol exposure to bioagents, and for holding animals for as long as several months after exposure to assess the safety and efficacy of candidate countermeasures.

Biocontainment facilities are complex, and the cost of designing, building, and operating them is high. NIAID will fund the construction of several additional BSL-3 laboratories and two BSL-4 laboratory research facilities that are expected to begin operation in the next 5 to 7 years (NIAID press release, 2003; Parker, 2003). These include additional intramural laboratories, one of which will be adjacent to USAMRIID, and extramural facilities. DHS (no date) is also planning the construction of a major facility for biodefense research and analysis at Fort Detrick. This facility is expected to include a biocontainment laboratory and aerosol exposure capabilities.

The existing major facilities at USAMRIID, however, are more than 30 years old and will not be useful much longer without extensive renovations. The estimated costs of replacing the USAMRIID facility to ensure the necessary scientific and physical capacity for modern research, testing, and evaluation are $1 billion over 8 years (USAMRIID, 2003). Although the Secretary of Defense has been given specific authority to use available construction funds to improve DoD laboratories that are necessary to carry out the department's research and development program for biomedical countermeasures,[8] the committee saw no indication that funds have been budgeted for the work needed at USAMRIID.

---

[8]National Defense Authorization Act for Fiscal Year 2004, P.L. 108-136, Sec. 1601 (November 2003).

It also is unclear whether sufficient facilities for holding, testing, and evaluation of animals in compliance with FDA regulations for Good Laboratory Practices[9] have been planned (Parker, 2003). Such facilities have to support compliance with stricter standards for consistency than are necessary during earlier stages of product research and development. They will be needed to conduct the animal testing required by the Animal Efficacy Rule. The committee is concerned that the limited availability of such facilities already constitutes a bottleneck for the testing and evaluation process (Linden, 2002; Parker, 2003; Peuschel, 2002) and that the backlog will grow more severe as additional discovery efforts bear fruit.

Another facilities issue is the capacity for cGMP production of candidate vaccines and drugs before full-scale manufacturing begins. Clinical testing of candidate products requires the use of material produced in compliance with FDA regulations for cGMP,[10] and cGMP production requires a sophisticated facility and skilled staff. Historically, vaccine and drug development has been the domain of industry; thus, licensed cGMP facilities, especially for vaccine production, are rare outside the industrial enterprise. For researchers who do not have an industry partner, lack of access to cGMP-compliant production facilities can be a major obstacle to clinical testing of candidate products (McCoy, 2003). Contract facilities are reported to have waiting times of 12 to 24 months (Shepard, 2003). This is a problem for developers of products to meet DoD needs, as well as for academic labs pursuing other niche products.

In view of the targeted market for medical countermeasures against biowarfare agents, industry may have limited incentives to make cGMP facilities available to manufacture the small amounts of material needed for early clinical testing. Increased access to facilities designed to produce materials for at least Phase 1 and 2 clinical trials would aid the development of candidate vaccines and drugs. Ideally, such facilities should have the flexibility to allow dedicated use of space for a variety of parallel products. For work on biodefense countermeasures, space meeting BSL-3 standards is also needed. Within DoD, the Walter Reed Army Institute of Research opened a facility in 1999 for production of cGMP-grade vaccines, with 25,000 square feet of space and five clean rooms (DoD, no date b). This facility has limited capacity compared with one opened in July 2003 by St. Jude Children's Research Hospital, which has a 64,000 square foot facility with 16 clean rooms (Personal communication, E. Tuomanen, St. Jude Children's Research Hospital, September 9, 2003).

---

[9]Good Laboratory Practice for Nonclinical Laboratory Studies (21 C.F.R. Pt. 58 [2003]).

[10]Current Good Manufacturing Practice in Manufacturing, Processing, Packing, or Holding of Drugs; General (21 C.F.R. Part 210 [2003]) and Current Good Manufacturing Practices for Finished Pharmaceuticals (21 C.F.R. Part 211 [2003]).

Because several different government agencies have research interests related to the development of biodefense products, interagency coordination is essential. A planning and assessment process is needed so that comprehensive estimates can be developed for all government agencies of biodefense requirements for laboratory space and for animal testing and holding space at BSL-3 and BSL-4 levels. This planning process should also take into account the anticipated demands of the academic community and industry. A high-level process is needed to prioritize the use of the limited space available and to continue planning for any additional space needs.

12. The Medical Biodefense Agency should participate in interdepartmental efforts to make a formal assessment of the need for facilities for animal testing and holding and for GMP-compliant manufacturing of material for clinical testing that will arise from research efforts to develop medical countermeasures to biowarfare or bioterrorism agents that are under way, planned, or likely.

13. The Medical Biodefense Agency should promote the development, and participate in a system for prioritizing the use, of specialized government-owned testing facilities that are essential for research and development of medical biodefense countermeasures.

14. DoD should provide funding to carry out the renovations necessary to ensure that USAMRIID can continue operation of fully functional BSL-3 and BSL-4 facilities for laboratory and animal research.

ENSURING THE AVAILABILITY OF A WELL-TRAINED WORKFORCE

A key to success in the research and development efforts on medical countermeasures against biowarfare agents is a well-trained and experienced workforce. People with relevant scientific, regulatory, and acquisition expertise are all integral to these efforts. At present, however, the supply of scientific and technical personnel is limited not only within DoD, but in the larger community of biowarfare and bioterrorism countermeasures development (IOM, 2002; NIAID, 2002; Partnership for Public Service, 2003; Top et al., 2000).

A recent report expressed concern about the small and shrinking federal workforce with medical and biological expertise relevant to responding to a biological attack, noting that many federal employees with this expertise are nearing retirement age and that limitations on pay, poor hiring procedures, and unattractive work settings make recruitment of replacements very difficult (Partnership for Public Service, 2003). Some of these problems have been reported as particularly acute within DoD (Defense Science Board, 1998, 2000, 2002) and were discussed with the committee during its information-gathering meetings.[11] Reports have also noted the limited expertise in biological sciences within DoD (Danzig and Berkowsky, 1997; NRC, 2001) and the apparent impact of the elimination of the military draft in the early 1970s on the numbers of researchers with medical and doctoral degrees coming through the military medical research laboratories (IOM, 2002; Top et al., 2000).

The lack of sufficient personnel with necessary expertise in infectious diseases, microbiology, immunology, primate medicine, drug development and vaccinology, and vaccine manufacturing is a limiting factor in the rate at which biomedical countermeasures can be developed. The committee notes in particular the need for people with expertise in aerobiology, in the development of animal models of human disease caused by biological warfare agents, and in the advanced development and manufacture of medical products, particularly vaccines. There is also a shortage of veterinarians with the necessary expertise in laboratory animal medicine, management, and pathology, including those with special training in the care and use of nonhuman primates (NRC, 2004).

In addition to personnel with these scientific and technical skills, DoD and other organizations involved in bringing medical products to licensure have a need for people who have a thorough understanding of the FDA regulatory process and, if possible, experience at FDA or in working with FDA to bring a product to licensure. Such regulatory expertise can inform the advanced development effort by helping to provide an understanding of the varying areas of flexibility and constraint in the regulations and of FDA's needs for data and documentation. Some have pointed specifically to DoD's need to improve its regulatory expertise in support of the medical biodefense program (Rettig and Brower, 2003). Clearly, FDA requires a highly trained staff as well, with incentives and support to maintain a rapid pace of efficient and thorough reviews.

Researchers and product developers within DoD also need an understanding of the defense acquisition system, the management process by which DoD oversees the development and production of products and technologies to meet defined needs. Famous for its complexity, the DoD acquisition system has undergone recent changes in policy with a goal of fostering efficiency, creativity, flexibility, and innovation (DoD, 2003a, b, c). Understanding how to work within this system requires training and

---

[11]See Appendix B for agendas for the committee's information-gathering meetings. Personnel issues were discussed at the January 2003 meeting.

experience. Such know-how is crucial for those who are managing development of medical countermeasures against biowarfare agents or naturally occurring infectious diseases. They must work not only within the law and regulations enforced by FDA, but also within DoD's rules for entering and progressing through its acquisition system, with milestones and evaluations to justify funding for continued development. Even though the committee advocates relying on the FDA regulatory process as the basis for technical judgments of progress in the development and performance of medical biodefense countermeasures, links with the DoD acquisition system must be maintained.

Several approaches are needed to address the serious workforce shortage in the biodefense area. The pipeline of young scientists in training to carry out the work of basic research and early development will be enhanced to some degree by recent research and training grants available through the NIAID biodefense research program (NIAID, 2003). However, DoD has certain unique resources in the form of scientists and technicians with aerobiology expertise and facilities suitable for aerosol exposures of laboratory animals. These unique resources should be used in training additional aerobiology investigators and technicians. DoD expertise could also contribute to the development of a curriculum for such training.

Further, the NIAID training program is unlikely to adequately address the critical need for people with experience in advanced development and manufacture of vaccines and drugs. Most of the expertise in vaccine development and manufacturing resides within the commercial sector, primarily in the four large vaccine-producing companies. The committee sees the need for additional training programs that involve the pharmaceutical industry as a resource and partner. A possible model for such a program exists in the National Institute of General Medical Sciences Biotechnology Training Program in which predoctoral trainees are expected to carry out an industrial internship (NIGMS, 2003).

15. The Medical Biodefense Agency should define the capabilities needed for its medical countermeasures workforce, collaborate with NIAID and industry to develop a training curriculum, and support training programs in the areas of special expertise needed for research and development of medical countermeasures. The Medical Biodefense Agency could contribute unique DoD resources in areas of aerobiology and the development of animal models of human diseases caused by biological warfare agents.

In addition to efforts to augment the supply of trained scientists, steps are needed to improve DoD's ability to hire and retain such scientists. The Medical Biodefense Agency will need the authority to offer competitive salaries and other incentives.

The National Defense Authorization Act for FY 2004 (P.L. 108-136) includes provisions that may meet some of these needs. Specifically, DoD received authority to enter into up to 30 personal services contracts, with compensation at rates above those for department employees, to carry out research and development activities for biomedical countermeasures. This authority can be used when the services to be provided are determined to be "urgent or unique" and if it is not practical to obtain those services in other ways. In addition, to accelerate research and development of biomedical countermeasures, the Secretary of Defense may hire scientific and technical personnel using special hiring authorities that include up to 5 years of supplemental payments in addition to salary. Overall, the department can hire up to 2,500 experts under such terms.

Use of mechanisms such as the Intergovernmental Personnel Act (5 U.S.C. §§ 3371–3375; 5 C.F.R. Part 334) should also be encouraged to enable the Medical Biodefense Agency to draw on expertise in other agencies and in academia and other not-for-profit organizations through temporary appointments. The agency should also explore mechanisms that could provide increased access to the expertise of industry employees.

16. DoD should use its authority under the National Defense Authorization Act for FY 2004 (P.L. 108-136) to offer more competitive salaries to technical experts to bring necessary expertise in biotechnology and pharmaceutical research and development to the Medical Biodefense Agency. Budgeting for the Medical Biodefense Agency should reflect the need to use such provisions to recruit experienced scientific and technical personnel.

REFERENCES

Albert V. 2003. Rapid development by Human Genome Sciences of a medical countermeasure to protect against anthrax. Presentation to the Institute of Medicine and National Research Council Committee on Accelerating the Research, Development, and Acquisition of Medical Countermeasures Against Biological Warfare Agents, Meeting IV. Washington, DC.

Aventis Pasteur. 2002. An Effective Action Plan for Anti Bioterrorism Vaccine Acquisition and Procurement. Position paper. Swiftwater, PA: Aventis Pasteur Inc.

Barbaro M. 2003. Biodefense plan greeted with caution. *Washington Post* (May 2), p. E1.

Bennett BT, Abee CR, Henrickson R (eds). 1995. *Nonhuman Primates in Biomedical Research. Biology and Management*. San Diego: Academic Press.

Black H. 2003. Monkey trouble. *The Scientist* (July 28). [Online]. Available: http://www.biomedcentral.com/news/20030728/02 [accessed July 28, 2003].

Carney JM. 2003. Accelerating research and development for biowarfare therapeutics. Presentation to the Institute of Medicine and National Research Council Committee on Accelerating the Research, Development, and Acquisition of Medical Countermeasures Against Biological Warfare Agents, Meeting II. Washington, DC.

CDC (Centers for Disease Control and Prevention). 2002. Possession, use, and transfer of select agents and toxins. Interim final rule. *Federal Register* 67(240):76886–76905.

Danzig R, Berkowsky PB. 1997. Why should we be concerned about biological warfare? *JAMA* 278(5):431–432.

Defense Science Board. 1998. *Report of the Defense Science Board Task Force on Defense Science and Technology Base for the 21st Century*. Washington, DC: Department of Defense, Office of the Under Secretary of Defense for Acquisition and Technology.

Defense Science Board. 2000. *Report of the Defense Science Board Task Force on the Technological Capabilities of Non-DoD Providers*. Washington, DC: Department of Defense, Office of the Under Secretary of Defense for Acquisition and Technology.

Defense Science Board. 2002. *2001 Summer Study on Defense Science and Technology*. Washington, DC: Office of the Under Secretary of Defense for Acquisition, Technology, and Logistics.

DeMarcus T. 2003. Nonhuman primate importation and quarantine, United States, FY2002. Centers for Disease Control and Prevention (slide presentation submitted to committee).

Department of Homeland Security. No date. Department of Homeland Security Budget in Brief. [Online]. Available: http://www.dhs.gov/interweb/assetlibrary/FY_2004_BUDGET_IN_BRIEF.pdf [accessed August 26, 2003].

Department of Homeland Security. 2003. Regulations implementing the Support Anti-Terrorism by Fostering Effective Technologies Act of 2002 (the SAFETY Act). Proposed rule. *Federal Register* 68(133):41420–41432.

DoD (Department of Defense). No date a. "Other Transaction." [Online]. Available: http://alpha.lmi.org/dodgars/other_transactions/other_transactions.htm [accessed May 20, 2003].

DoD. No date b. Department of Defense In-House RDT&E Activities: FY2000 Management Analysis Report. [Online]. Available: http://www.scitechweb.com/inhousereport/index1.html [accessed September 9, 2003].

DoD. 1993. *Department of Defense Directive: DoD Imminization Program for Biological Warfare Defense*. Number 6205.3. Washington, DC: Department of Defense.

DoD. 2000. *Department of Defense Directive: Use of Investigational New Drugs for Force Health Protection*. Number 6200.2. Washington, DC: Department of Defense.

DoD. 2001. "Other Transactions" (OT) Guide for Prototype Projects. Washington, DC: Under Secretary of Defense for Acquisition, Technology, and Logistics. [Online]. Available: http://www.acq.osd.mil/dp/dsps/ot/dspsot.htm [accessed September 2, 2003].

DoD. 2003a. *Department of Defense Directive: The Defense Acquisition System*. Number 5000.1. Washington, DC: Department of Defense.

DoD. 2003b. *Department of Defense Instruction: Operation of the Defense Acquisition System*. Number 5000.2. Washington, DC: Department of Defense.

DoD. 2003c. DoD New Acquisition Policies Instituted. News release, May 14. [Online]. Available: http://www.defenselink.mil/releases/2003/b05142003_bt327-03.html [accessed May 22, 2003].

FDA (Food and Drug Administration). 1990. Informed consent for human drugs and biologics; determination that informed consent is not feasible. Interim rule. *Federal Register* 55:52814.

FDA. 1998. Guidance for Industry: Fast Track Drug Development Programs—Designation, Development, and Application Review. U.S. Department of Health and Human Services, Food and Drug Administration, Center for Drug Evaluation and Research and Center for Biologics Evaluation and Research. [Online]. Available: http://www.fda.gov/cder/guidance/2112fnl.pdf [accessed August 11, 2003].

FDA. 2000. Anti-infective Drugs Advisory Committee (transcript of July 28 meeting). [Online]. Available: http://www.fda.gov/ohrms/dockets/ac/00/transcripts/3632t1a.pdf [accessed August 27, 2003].

FDA. 2001. Prescription drug products; doxycycline and penicillin G procaine administration for inhalational anthrax (post-exposure). Notice. *Federal Register* 66(213):55679–55682.

FDA. 2002a. Guidance for Industry: Inhalational Anthrax (Post-exposure)—Developing Antimicrobial Drugs. U.S. Department of Health and Human Services, Food and Drug Administration, Center for Drug Evaluation and Research. [Online]. Available: http://www.fda.gov/cder/guidance/4848dft.PDF [accessed August 11, 2003].

FDA. 2002b. New drug and biological drug products: Evidence needed to demonstrate effectiveness of new drugs when human efficacy studies are not ethical or feasible. Final rule. 21 C.F.R. Parts 314 and 601. *Federal Register* 67(105):37988–37998.

FDA. 2003a. FDA Approves Pyridostigmine Bromide as Pretreatment Against Nerve Gas. Press release, February 5. [Online]. Available: http://www.fda.gov/bbs/topics/NEWS/2003/NEW00870.html [accessed February 6, 2003].

FDA. 2003b. First Nasal Mist Flu Vaccine Approved. Press release, June 17. [Online]. Available: http://www.fda.gov/bbs/topics/NEWS/2003/NEW00913.html [accessed September 4, 2003].

FDA. 2003c. Pyridostigmine bromide package insert. [Online]. Available: http://www.fda.gov/cder/foi/label/2003/020414lbl.pdf [accessed September 24, 2003].

GAO (General Accounting Office). 2002. *Defense Acquisitions: DOD Has Implemented Section 845 Recommendations but Reporting Can Be Enhanced.* GAO-03-150. Washington, DC: U.S. General Accounting Office.

Geisbert TW, Pushko P, Anderson K, Smith J, Davis KJ, Jahrling PB. 2002. Evaluation in nonhuman primates of vaccines against Ebola virus. *Emerging Infectious Diseases* 8(5):503–507.

Gillis J. 2003. Md. firm plans human tests of anthrax antidote. *Washington Post* (March 18), p. A22.

Gottron F. 2003. *Project BioShield.* CRS Report for Congress, RS21507. Updated August 25. Washington, DC: Library of Congress, Congressional Research Service.

Hearn JP. 2003. Primate priorities—an international perspective. Pp. 3–9 in *International Perspectives: The Future of Nonhuman Primate Resources.* Washington, DC: National Academies Press.

Henchal EA. 2003. Medical product development at USAMRIID. Presentation to the Institute of Medicine and National Research Council Committee on Accelerating the Research, Development, and Acquisition of Medical Countermeasures Against Biological Warfare Agents, March 19. Fort Detrick, MD.

IOM (Institute of Medicine). 2002. *Protecting Our Forces: Improving Vaccine Acquisition and Availability in the U.S. Military.* Lemon SM, Thaul S, Fisseha S, O'Maonaigh HC, eds. Washington, DC: The National Academies Press.

Ivins BE, Fellows PF, Pitt MLM, Estep JE, Welkos SL, Worsham PL, Friedlander AM. 1996. Efficacy of a standard human anthrax vaccine against *Bacillus anthracis* aerosol spore challenge in rhesus monkeys. *Salisbury Medical Bulletin* 87(Suppl):125–126.

Ivins BE, Pitt ML, Fellows PF, Farchaus JW, Benner GE, Waag DM, Little SF, Anderson GW Jr., Gibbs PH, Friedlander AM. 1998. Comparative efficacy of experimental anthrax vaccine candidates against inhalation anthrax in rhesus macaques. *Vaccine* 16(11–12):1141–1148.

Jahrling PB. 2002. Medical countermeasures against the re-emergence of smallpox virus. Pp. 50–53 in *Biological Threats and Terrorism: Assessing the Science and Response Capabilities*, SL Knobler, AAF Mahmoud, and LA Pray, eds. Washington, DC: National Academy Press.

Linden C. 2002. Sponsor presentation on the study charge. Presentation to the Institute of Medicine and National Research Council Committee on Accelerating the Research, Development, and Acquisition of Medical Countermeasures Against Biological Warfare Agents, Meeting I. Washington, DC.

Linden C. 2003. Medical countermeasures for biological warfare agents research carried out at/through U.S. Army Medical Research and Materiel Command. Presentation to the Institute of Medicine and National Research Council Committee on Accelerating the Research, Development, and Acquisition of Medical Countermeasures Against Biological Warfare Agents, Meeting II. Washington, DC.

Ludlam C. 2003. Research, development, and procurement of countermeasures for bioterror attacks. Presentation to the Institute of Medicine and National Research Council Committee on Accelerating the Research, Development, and Acquisition of Medical Countermeasures Against Biological Warfare Agents, Meeting V. Washington, DC.

McCoy M. 2003. Serving emerging pharma. *Chemical & Engineering News* 81(21):21–33.

NIAID (National Institute of Allergy and Infectious Diseases). 2002. *NIAID Biodefense Research Agenda for CDC Category A Agents*. Bethesda, MD: National Institutes of Health. [Online]. Available: http://www.niaid.nih.gov/biodefense/research/biotresearch agenda.pdf [accessed July 1, 2002].

NIAID. 2003. Biodefense Research Training and Career Development Opportunities. Notice: NOT-AI-03-047. Bethesda, MD.

NIAID and USAMRIID (U.S. Army Medical Research Institute of Infectious Diseases). 2002. Interagency agreement between the National Institute of Allergy and Infectious Diseases and the U.S. Army Medical Research Institute of Infectious Diseases.

NIGMS (National Institute of General Medical Sciences). 2003. Biotechnology Predoctoral Training Program. [Online]. Available: http://www.nigms.nih.gov/funding/trngmech.html [accessed July 30, 2003].

NRC (National Research Council). 2001. *Opportunities in Biotechnology for Future Army Applications*. Washington, DC: National Academy Press.

NRC. 2003. *International Perspectives: The Future of Nonhuman Primate Resources, Proceedings of the Workshop held April 17-19, 2002*. Washington, DC: The National Academies Press.

NRC. 2004. *National Need and Priorities for Veterinarians in Biomedical Research*. Washington, DC: The National Academies Press.

Parker G. 2003. Biocontainment facilities needed to support medical biological defense research, development, test, and evaluation. Presentation to the Institute of Medicine and National Research Council Committee on Accelerating the Research, Development, and Acquisition of Medical Countermeasures Against Biological Warfare Agents, Meeting IV. Washington, DC.

Partnership for Public Service. 2003. *Homeland Insecurity: Building the Expertise to Defend America from Bioterrorism*. Washington, DC: Partnership for Public Service. [Online]. Available: http://www.ourpublicservice.org/publications3735/publications_show.htm?doc_id=181630 [accessed July 8, 2003].

Peuschel M. 2002. Army, NIH team up to counter bioterror. *U.S. Medicine* (November). [Online]. Available: http://www.usmedicine.com/article.cfm?articleID=533&issueID=44 [accessed August 16, 2003].

Pitt MLM, Ivins BE, Estep JE, Farchaus J, Friedlander AM. 1996. Comparison of the efficacy of purified protective antigen and MDPH to protect non-human primates from inhalation anthrax. *Salisbury Medical Bulletin* 87(Suppl):130.

Pitt ML, Little SF, Ivins BE, Fellows P, Barth J, Hewetson J, Gibbs P, Dertzbaugh M, Friedlander AM. 2001. In vitro correlate of immunity in a rabbit model of inhalational anthrax. *Vaccine* 19(32):4768–4773.

Read L. 2003. Furthering public health security: Project Bioshield. Testimony before the Subcommittee on Health of the House Committee on Energy and Commerce and the Subcommittee on Emergency Preparedness and Response of the House Committee on Homeland Security, March 27. [Online]. Available: http://energycommerce.house.gov/108/Hearings/03272003hearing844/Read1393print.htm [accessed October 15, 2003].

Research Resources Information Center. No date. Rhesus Monkey Demands in Biomedical Research. A Workshop Report. Based on a workshop held April 19–20, 2002. Bethesda, MD.

Rettig R. 1999. *Military Use of Drugs Not Yet Approved by the FDA for CW/BW Defense: Lessons from the Gulf War.* MR-1019/9-OSD. Santa Monica, CA: RAND Institute. [Online]. Available: http://www.rand.org/publications/MR/MR1018.9/ [accessed February 10, 2003].

Rettig R, Brower J. 2003. *The Acquisition of Drugs and Biologics for Chemical and Biological Warfare Defense: Department of Defense Interactions with the Food and Drug Administration.* Santa Monica, CA: RAND Institute. [Online]. Available: http://www.rand.org/publications/MR/MR1659/ [accessed October 20, 2003].

Ryan U. 2003. Project BioShield Act of 2003. Testimony before the House Government Reform Committee, April 4. Washington, DC.

Shepard S. 2003. A new level of research. *Memphis Business Journal* (July 14). [Online]. Available: http://memphis.bizjournals.com/memphis/stories/2003/07/14/focus1.html [accessed July 18, 2003].

Shortage of monkeys hampering research. 2003. *Seattle Times* (August 10), p. A8.

Sullivan NJ, Geisbert TW, Beisbert JB, Xu L, Yang Z, Roederer M, Koup RA, Jahrling PB, Nabel GJ. 2003. Accelerated vaccination for Ebola virus haemorrhagic fever in non-human primates. *Nature* 424:681–684.

Tether T. 2002. Statement submitted to the Subcommittee on Technology and Procurement Policy, Committee on Government Reform, United States House of Representatives, May 10. Washington, DC [Online]. Available: http://reform.house.gov/tapps/hearings/5-10-02/DARPAtestimonyfinal.htm [accessed May 19, 2003].

Top FH Jr., Dingerdissen JJ, Habig WH, Quinnan GV Jr., Wells RL. 2000. DoD Acquisition of Vaccine Production: Report to the Deputy Secretary of Defense by the Independent Panel of Experts. In DoD. 2001. *Report on Biological Warfare Defense Vaccine Research and Development Programs.* Washington, DC: Department of Defense. [Online]. Available: http://www.acq.osd.mil/cp/bwdvrdp-july01.pdf [accessed February 19, 2004].

USAMRIID (U.S. Army Medical Research Institute of Infectious Diseases). 2003. An Update Report to the Congress Concerning the U.S. Army Medical Research Institute of Infectious Diseases (USAMRIID) Feasibility Study and Biodefense Campus at Fort Detrick, Maryland. Submitted in Compliance with House Report 107-350. September. Fort Detrick, MD: U.S. Army Medical Research and Materiel Command.

Zaucha GM, Jahrling PB, Geisbert TW, Swearengen JR, Hensley L. 2001. The pathology of experimental aerosolized monkeypox virus infection in cynomolgus monkeys (*Macaca fascicularis*). *Laboratory Investigation* 81(12):1581–1600.

APPENDIX A

Background: The Current DoD Medical Biowarfare Countermeasures Program

This appendix provides background information on medical biodefense activities in the Department of Defense (DoD). The DoD components involved in planning, managing, and executing these activities are identified first, followed by a summary of the legislative mandates and DoD policies that guide work on medical countermeasures. The identification of biological agents considered to pose a threat is briefly described. The appendix concludes with information on medical biodefense countermeasures that are currently available and under development.

ORGANIZATION OF DOD MEDICAL BIOLOGICAL DEFENSE ACTIVITIES

As described in part in Chapters 1 and 2, the organization of DoD's program for developing medical countermeasures against biological warfare agents reflects a 1993 congressional mandate (P.L. 103-160; 50 U.S.C. 1522) that all of DoD's chemical and biological defense activities, both medical and nonmedical, be coordinated by a single office within the Office of the Secretary of Defense and managed through the DoD Acquisition Board process.

The current configuration of responsibilities for the DoD Chemical and Biological Defense Program is described in an implementation plan issued in April 2003 (Aldridge, 2003) (see Figure 2-1). Responsibility for chemical and biological defense activities falls under the purview of the Under Secretary of Defense for Acquisition, Technology, and Logistics

(USD(AT&L)), who also serves as the senior acquisition official. The responsibility for coordination and integration of the Chemical and Biological Defense Program is assigned to the Assistant to the Secretary of Defense for Nuclear and Chemical and Biological Defense Programs (ATSD(NCB)) and exercised by the Deputy ATSD for Chemical and Biological Defense (DATSD(CBD)).

The chemical and biological defense requirements and priorities of the combatant forces from all of the military services guide program planning and budgeting for the Chemical and Biological Defense Program. Those requirements and priorities are developed and managed by the Joint Requirements Office for Chemical, Biological, Radiological, and Nuclear Defense (JRO-CBRN), which was chartered in February 2003 (Pace, 2003). The responsibilities of the JRO-CBRN include coordination with the intelligence community in the development of threat assessments. The JRO-CBRN reports to the Chairman of the Joint Chiefs of Staff.

The Army is designated as the executive agent for the Chemical and Biological Defense Program, with responsibility for coordinating and integrating research, development, test and evaluation, and acquisition requirements for the military services. Management and execution of these activities are carried out by other organizations. Responsibility for managing the intramural and extramural research activities (the science and technology base) of the Chemical and Biological Defense Program is assigned to the Defense Threat Reduction Agency (DTRA), which reports to the USD(AT&L). For medical biodefense countermeasures, intramural science and technology activities are carried out primarily through the U.S. Army Medical Research Institute of Infectious Diseases (USAMRIID), which is part of the U.S. Army Medical Research and Materiel Command. Some intramural medical biodefense work is carried out at other military laboratories as well.

Responsibility for advanced development and acquisition of chemical and biological countermeasures is assigned to the Joint Program Executive Office for Chemical and Biological Defense Programs (JPEO-CBD). Within the JPEO-CBD, medical biodefense countermeasures are the responsibility of the Chemical Biological Medical Systems (CBMS) office, which includes the Joint Vaccine Acquisition Program (JVAP) and Medical Identification and Treatment Systems (covering therapeutic drugs and diagnostics). All vaccine development is being handled through a prime systems contract, awarded in 1997 to DynPort Vaccine Company LLC (DVC). As a prime systems contractor, DVC manages the advanced development of vaccine candidates through contracts with various companies to perform tasks involved in developing, testing, and delivering the vaccines. DVC does not have laboratory or vaccine production facilities of its own.

The Defense Advanced Research Projects Agency (DARPA) is authorized to conduct a separate extramural program of basic and applied research in chemical and biological defense and, over the past 5 years, has included work on medical biodefense countermeasures. DARPA was established in 1958 to permit more flexible approaches to long-horizon, high-risk, high-payoff research within DoD. Its research program is shaped, in part, by the expertise and interests of program managers, who are recruited for 4-year periods. Programs do not necessarily continue beyond the tenure of their program managers. Some of the promising DARPA-funded projects on medical countermeasures have been transferred to the science and technology base of the Chemical and Biological Defense Program for further work.

DOD POLICIES RELATED TO MEDICAL DEFENSE AGAINST BIOLOGICAL WARFARE

Two directives establish DoD policies and requirements related to medical aspects of biological warfare defense. Directive 6205.3, "DoD Immunization Program for Biological Warfare Defense," specifies that personnel assigned or scheduled for deployment to a high-threat area should be immunized against validated biological warfare threat agents for which suitable vaccines are available (DoD, 1993). The priorities for vaccine research and development are identified as including the development of vaccines against threat agents for which no vaccines exist, the improvement of vaccines that are slow to produce immunity or require multiple doses, and the development of multivalent vaccines. These vaccines are to be either licensed by the Food and Drug Administration (FDA) or designated for use as Investigational New Drugs (INDs). The directive also calls for validating and prioritizing the biological warfare threats annually.

Directive 6200.2, "Use of Investigational New Drugs for Force Health Protection," establishes DoD policies and procedures for use of IND countermeasures when no FDA-approved product is available (DoD, 2000), as required by 10 U.S.C. 1107 and Executive Order 13139 (Clinton, 1999). Use of FDA-approved products is preferred, but when they are not available, DoD components may request approval of the Secretary of Defense to use an IND product. The request must be justified on the basis of the available evidence of the safety and efficacy of the drug or vaccine and the nature and degree of the threat. DoD must then develop a protocol for use of the IND product, including providing for informed consent of service members before they receive it. Only the President may grant a waiver of informed consent, in response to a request from the Secretary of Defense.

Relevant aspects of DoD policy, as summarized in a recent report (DoD, 2002), are shown in Box A-1.

BOX A-1
Key Features of DoD Policy and Requirements Concerning Acquisition and Use of Medical Countermeasures to Protect the Health of Military Forces Against Biological Warfare Threat Agents

■ Use FDA-licensed, commercially available medical countermeasures (i.e., vaccines) to protect the health of U.S. forces from biological warfare threat agents.

■ Employ IND medical countermeasures only when FDA-licensed products are unavailable and

♦ There is a confirmed high risk to force health protection that necessitates consideration of IND product use;

♦ Only after an in-depth review and approval by the Secretary of Defense of a request initiated by a Commander of a Combatant Command through the Chairman of the Joint Chiefs of Staff and in coordination with the ASD(HA) (Assistant Secretary of Defense for Health Affairs), the Under Secretary of Defense for Policy, the Secretary of the Army as the Executive Agent, and the DoD General Counsel;

♦ In strict compliance with a specific treatment protocol developed for the required indication that has been reviewed by the FDA and complies with the requirements of 21 C.F.R., including the requirement for informed consent; and

♦ Informed consent may be waived only by the President upon request by the Secretary of Defense and only under specified conditions; if a presidential waiver of informed consent is approved, then

• DoD Components must conduct ongoing monitoring and adhere to periodic reporting as required by the President,

• The Secretary of Defense shall notify Congress, and the public by Federal Register notification, as soon as practicable,

• The Secretary of Defense shall notify the President and the FDA Commissioner of any changed circumstances concerning the need to waive informed consent, and

• The waiver of informed consent terminates after 1 year or when no longer needed—whichever is earlier.

■ Vaccinate "at-risk" personnel against validated biological warfare threat agents in sufficient time for them to develop immunity before deployment to high-threat areas.

■ Integrate and prioritize efforts for vaccine research, development, testing, evaluation, acquisition, and stockpiling and to improve existing vaccines against validated biological warfare threat agents.

■ Develop a capability to acquire and stockpile adequate quantities of vaccines to protect the programmed force against all validated biological warfare threats.

SOURCE: DoD, 2002, p. 7.

IDENTIFICATION OF BIOLOGICAL THREAT AGENTS

Each year DoD generates a list of the biological warfare agents believed to pose the highest risks to military forces. The military services, through the combatant commanders, provide Joint Chiefs of Staff with their assessment of the biological warfare threats within their respective theaters. These assessments are validated in consultation with the intelligence community and result in a classified "threat list."

The DoD Chemical and Biological Defense Program Annual Report to Congress provides an unclassified discussion of this threat (e.g., DoD, 2003). The 2003 report notes that chemical and biological weapons are generally easier to develop, hide, and deploy than nuclear weapons and could be readily available to those with the will and resources to obtain them. According to the report, a dozen countries are believed to have biowarfare programs. Terrorist groups are also reported to be interested in these weapons, and the proliferation of chemical and biological weapons is expected to continue. An unclassified list of potential biological warfare threats is provided in Box A-2.

The Centers for Disease Control and Prevention (CDC, no date) and the National Institute of Allergy and Infectious Diseases (NIAID, no date) have published lists of bioterrorism threats (see Box A-3). These agents are classified as Category A, B, or C hazards on the basis of the readiness with which they can be disseminated, projected mortality or morbidity rates, and the need for special actions for public health preparedness. The highest priority is addressing the threats posed by Category A agents. The CDC and NIAID lists are very similar to the lists of threats of concern to DoD (Box A-2).

MEDICAL BIODEFENSE COUNTERMEASURES

Countermeasures Currently Available

The few FDA-approved countermeasures available against potential biological threat agents identified by DoD in its unclassified lists or against CDC's Category A agents are listed in Table A-1. Licensed vaccines exist for both anthrax and smallpox, and the military has access to adequate supplies of both. However, both vaccines pose substantial challenges. As licensed by FDA, the anthrax vaccine (BioThrax) is to be administered in six doses over 18 months.[1] The smallpox vaccine (Dryvax) requires only

[1]CDC is carrying out a clinical trial with associated animal studies to evaluate the safety and efficacy of a reduced number of doses and a different route of administration for this vaccine.

BOX A-2
Potential Biological Threats as Presented by DoD Medical Biological Defense Research and Development Program

Bacteria:
Bacillus anthracis (anthrax)
Yersinia pestis (plague)
Francisella tularensis (tularemia)
Brucella sp. (brucellosis)
Burkholderia maellei (glanders)
Coxiella burnetii (Q fever)

Toxins:
Botulinum toxins (types A–G)
Staphylococcal enterotoxins (SEA/B)
Ricin toxin
Marine neurotoxins
Mycotoxins
Clostridium perfringens

Viruses:
Smallpox
Encephalomyelitis viruses
Ebola
Marburg

SOURCE: Skvorak, 2003.

one dose, but brings with it concerns resulting from historical rates of severe side effects. Recent military experience with Dryvax suggests the occurrence of moderate or serious side effects in 0.1 percent of recipients (Grabenstein and Winkenwerder, 2003). Recent experience in the civilian community indicates that serious adverse events have been reported in 0.3 percent of recipients (CDC, 2003). The military as well as the public health community seek improved vaccines against both of these biowarfare agents. No licensed vaccines are available against botulism, plague, tularemia, or the viral hemorrhagic fevers, although vaccines against all of these diseases are under development.

Table A-1 also notes the availability of drugs with approved indications for the treatment of disease resulting from several of the threat agents of concern to DoD and the nation. Three antibiotics each have been approved for use in treating anthrax, plague, and tularemia. No therapeutics have been approved for use against botulism or the viruses of greatest concern. Table A-1 also notes the availability of medical countermeasures that have reached IND status and thus might be available for contingency use.

BOX A-3
Diseases and Biological Agents Identified by CDC as Posing a Threat to National Security

Category A
Anthrax (*Bacillus anthracis*)
Botulism (*Clostridium botulinum toxin*)
Plague (*Yersinia pestis*)
Smallpox (variola major)
Tularemia (*Francisella tularensis*)
Viral hemorrhagic fevers
(filoviruses [e.g., Ebola, Marburg] and arenaviruses [e.g., Lassa, Machupo])

Category B
Brucellosis (*Brucella* species)
Epsilon toxin of *Clostridium perfringens*
Food safety threats
(e.g., *Salmonella* species, *Escherichia coli* O157:H7, *Shigella*)
Glanders (*Burkholderia mallei*)
Melioidosis (*Burkholderia pseudomallei*)
Psittacosis (*Chlamydia psittaci*)
Q fever (*Coxiella burnetii*)
Ricin toxin from castor beans (*Ricinus communis*)
Staphylococcal enterotoxin B
Typhus fever (*Rickettsia prowazekii*)
Viral encephalitis
(alphaviruses [e.g., Venezuelan equine encephalitis, eastern equine encephalitis,western equine encephalitis])
Water safety threats
(e.g., *Vibrio cholerae, Cryptosporidium parvum*)

Category C
Emerging infectious diseases such as Nipah virus and hantavirus

SOURCE: CDC, no date.

Countermeasures Under Development

With so few FDA-approved countermeasures available against the highest-priority biological threats, the list of countermeasures that are needed is long. Table A-2 identifies the medical countermeasures that are in various stages of research and development, including six vaccines and

TABLE A-1 Medical Countermeasures Available Against Biological Threat Agents

Disease (Agent)	Licensed Vaccines	Approved Therapeutics	INDs for Contingency Use	Additional Information
Anthrax (*Bacillus anthracis*)	Yes (Anthrax Vaccine Adsorbed [BioThrax])	Doxycycline Ciprofloxacin Penicillin	BioThrax for postexposure use	Vaccine is licensed for 6 doses over 18 months, logistically difficult for military use CDC study under way to seek support for labeling change for different route of administration, use of fewer doses
Botulism (*Clostridium botulinum* toxin)	No	No	Botulinum toxoid vaccine Botulinum immune globulins (human and equine)	
Plague (*Yersinia pestis*)	No	Gentamicin Doxycycline Ciprofloxacin	None	

Smallpox (variola major)	Yes (Dryvax)	No	Vaccinia immune globulin Cidofovir Dryvax (postexposure) Aventis Pasteur vaccine (1:5 dilution)	Risk of adverse events from vaccine use, including mortality, is higher than for many other vaccines
Tularemia (*Francisella tularensis*)	No	Gentamicin Ciprofloxacin Doxycycline	LVS vaccine	
Encephalitis (equine encephalitis viruses)	No	No		
Venezuelan			Live and inactivated vaccines	
Eastern			Inactivated vaccine	
Western			Inactivated vaccine	

NOTE: LVS, live vaccine strain.
SOURCE: Clayson, 2003

TABLE A-2 DoD-Related Research and Development Activities for Medical Biological Warfare Countermeasures

Disease (Agent)	Countermeasure	Status	Projected Licensure
Smallpox (variola major)	Cell-cultured smallpox vaccine (CCSV)	DVC Phase 1 testing completed; "down select" decision between DoD and NIH candidates expected November 2003 from Defense Science Board	
	Vaccinia immune globulin (VIG)	DVC: Two Phase 1 trials on liquid product completed July 2003 Pivotal clinical trial on lyophilized product completed October 2000 Fast-track approval granted by FDA, final BLA submission anticipated 2004	FY 2005
	Intravenous cidofovir	Ongoing research at USAMRIID	
	Oral therapeutic antiviral drugs based on cidofovir, or on non-DNA polymerase target	Ongoing research at USAMRIID	
Anthrax (*Bacillus anthracis*)	Recombinant anthrax vaccine (rPA, derived from (*Escherichia coli*)	DVC Phase 1 trials begun October 2002; clinical trials will be completed but future work now on hold due to lack of funding	
	Recombinant anthrax vaccine (rPA, derived from *B. anthracis*, developed at USAMRIID)	To undergo Phase 1 trials funded by NIAID	

TABLE A-2 Continued

Disease (Agent)	Countermeasure	Status	Projected Licensure
Botulism (*Clostridium botulinum* toxin)	Recombinant bivalent botulinum vaccine (against neurotoxin serotypes A and B)	DVC submission of IND and start of Phase 1 trial anticipated in first half of 2004	FY 2012
	Pentavalent botulinum vaccine (against neurotoxin serotypes A, B, C, E, and F)	DVC discovery and preclinical development, to be funded by NIAID	
	Heptavalent botulinum vaccine (against neurotoxin serotypes A, B, C, D, E, F, and G)	DVC discovery and preclinical development, to be funded by NIAID	
Plague (*Yersinia pestis*)	Recombinant plague vaccine (F1-V fusion antigen)	DVC: Beginning of cGMP manufacturing anticipated in 2003 Phase 1 trial start anticipated for February 2005	FY 2014
	U.K. Recombinant plague vaccine (F1 + V mixture)	DVC planning Phase 1 trial in 2004	
Tularemia (*Francisella tularensis*)	LVS vaccine	DVC funded by DoD through FY 2003; funded by NIAID for Phase 1 trial	
Venezuelan equine encephalitis	Vaccine	DVC: IND submission anticipated by March 2004 Phase 1 trial start anticipated June 2004	
Ricin toxin	Recombinant ricin vaccine	Ongoing research at USAMRIID	
	Therapeutics for exposure to ricin	Ongoing research at USAMRIID	

Continued

TABLE A-2 Continued

Disease (Agent)	Countermeasure	Status	Projected Licensure
Brucellosis (*Brucella* sp.)	Vaccine candidate: orally administered MNPH1 live attenuated deletion mutant	Ongoing research at WRAIR	
Staphylococcal enterotoxin (SE) A and B	Vaccine candidates	Ready for transition to advanced developer USAMRIID scientists conducting stability analysis on pilot lots for use in future clinical studies	
	Therapeutics for exposure to SEs	Ongoing research at USAMRIID	
Glanders (*Burkholderia mallei*)	Vaccine candidates	Ongoing research at USAMRIID	
Viral hemorrhagic fevers (filoviruses, including Ebola and Marburg)	Multiagent vaccine capable of protecting against various Ebola and Marburg viruses	Ongoing research at USAMRIID	
	Antivirals	Ongoing research at USAMRIID	
	Immunotherapies for filoviruses	Ongoing research at USAMRIID	
General	Alternate vaccine delivery methods (oral, respiratory, transdermal) and adjuvants	Ongoing research at USAMRIID	

NOTE: BLA, Biologics License Application; cGMP, current Good Manufacturing Practice; LVS, live vaccine strain, U.K., United Kingdom; USAMRIID, U.S. Army Medical Research Institute of Infectious Diseases; WRAIR, Walter Reed Army Institute of Research.
SOURCES: DVC, 2003a, b, c; Skvorak, 2003; Personal communication, T. Irgens, DVC, September 16, 2003; November 12, 2003.

one immune globulin product in clinical testing under the auspices of the DoD Chemical and Biological Defense Program.

Until recently, basic research and early development of candidate countermeasures took place primarily through DoD. The substantial funding that became available to NIAID in fiscal year 2003 for the research and development of bioterrorism countermeasures has greatly increased the number and variety of medical countermeasure candidates under investigation. Information on the NIAID research agenda to develop countermeasures against Category A agents is available from the NIAID website (http://www.niaid.nih.gov/biodefense/) and a recent progress report (NIAID, 2003).

REFERENCES

Aldridge EC. 2003. Memorandum: Implementation Plan for the Management of the Chemical Biological Defense Program (CBDP). Washington, DC: Department of Defense.

CDC (Centers for Disease Control and Prevention). No date. Biological Diseases/Agents List. [Online]. Available: http://www.bt.cdc.gov/Agent/agentlist-category.asp [accessed November 4, 2002].

CDC. 2003. Update: adverse events following civilian smallpox vaccination—United States, 2003. *MMWR* 52(34):819–820.

Clayson JP. 2003. Joint Vaccine Acquisition Program (JVAP) Overview. Presentation to the Armed Forces Epidemiological Board, May 20. [Online]. Available: http://www.ha.osd.mil/afeb/meeting/052003meeting/default.cfm [accessed August 28, 2003].

Clinton WJ. 1999. Improving Health Protection of Military Personnel Participating in Particular Military Operations. Executive order 13139 of September 30, 1999. *Federal Register* 64(192):54175–54178.

DoD (Department of Defense). 1993. Department of Defense Directive: DoD Immunization Program for Biological Warfare Defense. Number 6205.3. Washington, DC: Department of Defense.

DoD. 2000. Department of Defense Directive: Use of Investigational New Drugs for Force Health Protection. Number 6200.2. Washington, DC: Department of Defense.

DoD. 2002. Acceleration of Research, Development, and Production of Medical Countermeasures for Defense Against Biological Warfare Agents. Report to Congress required by P.L.107-107, Section 1044(e). Washington, DC: Department of Defense.

DoD. 2003. Department of Defense Chemical and Biological Defense Program. Volume I: Annual Report to Congress. Washington, DC: Department of Defense. [Online]. Available: http://www.acq.osd.mil/cp/nbc03/vol1-2003cbdpannualreport.pdf [accessed June 27, 2003].

DVC (DynPort Vaccine Company). 2003a. The pipeline: DVC products and technical support. *Vaccine Technology and Development News* 2(6):2. [Online]. Available: http://www.dynport.com/press.htm [accessed August 18, 2003].

DVC. 2003b. Press release: CSC Joint Venture Completes Phase I Vaccinia Immune Globulin Clinical Trial, July 31. [Online]. Available: http://www.dynport.com/press.htm [accessed August 18, 2003].

DVC. 2003c. Press release: CSC–Porton Joint Venture Wins $11 Million Award to Expedite Production of Botulinum Vaccine, September 4. [Online]. Available: http://www.dynport.com/press.htm [accessed September 8, 2003].

Grabenstein JG, Winkenwerder W. 2003. US military smallpox vaccination program experience. *JAMA* 289(24):3278–3282.

NIAID (National Institute of Allergy and Infectious Diseases). No date. NIAID Category A, B, & C Priority Pathogens. [Online]. Available: http://www.niaid.nih.gov/biodefense/bandc%5Fpriority.htm [accessed September 24, 2003].

NIAID. 2003. NIAID Biodefense Research Agenda for CDC Category A Agents: Progress Report. Bethesda, MD: National Institutes of Health. [Online]. Available: http://www.niaid.nih.gov/biodefense/research/category_A_Progress_Report.pdf [accessed August 11, 2003].

Pace P. 2003. Charter for the Joint Requirements Office for Chemical, Biological, Radiological and Nuclear Defense [memorandum and enclosure]. Washington, DC: Department of Defense, Joint Chiefs of Staff.

Skvorak JP. 2003. Medical Chemical Biological Defense Research Program. U.S. Army Medical Research and Materiel Command. Presentation to the Armed Forces Epidemiological Board, May 20. [Online]. Available: http://www.ha.osd.mil/afeb/meeting/052003meeting/default.cfm [accessed August 28, 2003].

APPENDIX B

Agendas for Information-Gathering Meetings

MEETING I
DECEMBER 16–17, 2002

The National Academy of Sciences Building
2101 Constitution Avenue, N.W.
Washington, DC

Monday, December 16, 2002

1:15 p.m.	Introductory Remarks *Leslie Benet, Ph.D., Chair, Committee on Accelerating the Research, Development, and Acquisition of Medical Countermeasures Against Biological Warfare Agents* Introductions by Committee Members and Meeting Attendees
1:30	Sponsor Presentation on the Study Charge *Carol Linden, Ph.D., Director, Medical Chemical and Biological Defense Research Program, U.S. Army Medical Research and Materiel Command* Congressional Comment on the Study Charge *Mr. Jean Reed, Professional Staff Member, House Armed Services Committee* Discussion

2:30 Status of Department of Defense Effort on Accelerating Research, Development, and Acquisition of Medical Countermeasures
Anna Johnson-Winegar, Ph.D., Deputy Assistant to the Secretary of Defense for Chemical and Biological Defense

Discussion

3:15 Break

3:30 CBER Perspective on Opportunities for Accelerating Licensure of Medical Countermeasures for Biowarfare Agents
Kathryn Zoon, Ph.D., Director, Center for Biologics Evaluation and Research, Food and Drug Administration

Discussion

4:15 CDER Perspective on Opportunities for Accelerating Licensure of Medical Countermeasures for Biowarfare Agents
Janet Woodcock, M.D., Director, Center for Drug Evaluation and Research, Food and Drug Administration

Discussion

5:00 Adjourn

Tuesday, December 17, 2002

8:00 a.m. Breakfast

8:30 IOM Report, September 2002: *Protecting Our Forces: Improving Vaccine Acquisition and Availability in the U.S. Military*
Stanley Lemon, M.D., Chair, IOM Committee on a Strategy for Minimizing the Impact of Naturally Occurring Infectious Diseases of Military Importance: Vaccine Issues in the U.S. Military

Discussion

9:15 DoD Acquisition of Vaccine Production
Report to the Deputy Secretary of Defense, December 2000
Gerald V. Quinnan, Jr., M.D., Professor and Chair of Preventive Medicine, Uniformed Services University of the Health Sciences

Discussion

10:00 Adjourn

MEETING II
JANUARY 14–15, 2003

The National Academies Keck Center
500 Fifth Street, N.W.
Washington, DC

Tuesday, January 14, 2003

9:00 a.m. Breakfast

9:30 Introductory Remarks
Leslie Benet, Ph.D., Chair, Committee on Accelerating the Research, Development, and Acquisition of Medical Countermeasures Against Biological Warfare Agents

9:45 Medical Countermeasures for Biological Warfare Agents
Research Carried out at/through the U.S. Army Medical Research and Materiel Command (USAMRMC)
Carol Linden, Ph.D., Director, Medical Chemical and Biological Defense Research Program, USAMRMC

10:30 Break

10:45 Chemical Biological Medical Systems (CBMS) (formerly Joint Vaccine Acquisition Program)
COL David Danley, Ph.D., Director, CBMS

11:30 DynPort Vaccine Company LLC
Michael J. Langford, D.V.M., Ph.D., Senior Vice President and Chief Scientific Officer, DynPort Vaccine Company LLC

12:00 p.m.	Working Lunch
12:45	Defense Threat Reduction Agency (DTRA) *LTC Keith R. Vesely, D.V.M., Ph.D., Chief, Medical Science and Technology Division, Chemical and Biological Defense Directorate*
1:15	Defense Advanced Research Projects Agency (DARPA) *John Carney, Ph.D., Unconventional Pathogen Countermeasures Program*
1:45	Joint Program Executive Office for Chemical and Biological Defense (JPEO-CBD) *BG Stephen Reeves, Program Executive Officer, Chemical and Biological Defense*
2:15	Joint Requirements Office for Chemical, Biological, Radiological, and Nuclear (JRO-CBRN) Defense *MAJ Ronald Fizer, Joint Requirements Integrator, JRO-CBRN Defense*
2:45	Break
3:00	Directed Roundtable Discussion: Representatives of Involved Government and Other Organizations

3:00	I	Basic Research Early Development
3:45	II	Transition to Advanced Development
4:30	III	Advanced Development Licensure

5:30	Summary and General Discussion
6:00	Adjourn

MEETING III
MARCH 17–18, 2003

The National Academies Keck Center
500 Fifth Street, N.W.
Washington, DC

Monday, March 17, 2003

9:00 a.m. Breakfast

9:30 Introductory Remarks
Leslie Z. Benet, Ph.D., Chair, Committee on Accelerating the Research, Development, and Acquisition of Medical Countermeasures Against Biological Warfare Agents

9:45 Current and Planned Activities of the National Institute of Allergy and Infectious Diseases (NIAID) in Research and Development of Medical Countermeasures Against Biological Warfare Agents
Anthony S. Fauci, M.D., Director, NIAID

Discussion

11:30 Break

11:45 Working Lunch

12:30 p.m. Opportunities for the Food and Drug Administration (FDA) to Contribute to Accelerating the Development and Licensure of Medical Countermeasures Against Biological Warfare Agents
Mark B. McClellan, M.D., Ph.D., Commissioner, FDA

Discussion

2:00 Break

2:15 Roundtable Discussion: Issues Related to the FDA Role in Development and Licensure of Medical Countermeasures Against Biological Warfare Agents

Maria Bernwinkler, Director, Clinical Research, PPD
LTC Robert Borowski, Ph.D., Senior Medical Advisor, Office of the Deputy to the Assistant to the Secretary of Defense for Chemical and Biological Defense, DoD
COL David Danley, Ph.D., Director, Chemical and Biological Medical Systems, DoD
Karen Goldenthal, M.D., Director, Division of Vaccines and Related Products Applications, Center for Biologics Evaluation and Research, FDA
Carole Heilman, Ph.D., Director, Division of Microbiology and Infectious Diseases, NIAID, NIH
COL Charles Hoke, Jr., M.D., U.S. Army Medical Research and Materiel Command (USAMRMC), DoD
Virginia Johnson, Ph.D., Deputy Product Development Manager, DynPort Vaccine Company LLC
Carol Linden, Ph.D., Director, Medical Chemical and Biological Defense Research Program, USAMRMC, DoD
John McCormick, M.D., Orphan Products Development, FDA
Andrea Meyerhoff, M.D., Director, Counterterrorism Programs, Office of the Commissioner, FDA
Dianne Murphy, M.D., Director, Office of Counterterrorism and Pediatric Drug Development, Center for Drug Evaluation and Research, FDA
Jerald Sadoff, M.D., Washington, DC

4:30 Adjourn Open Session

Tuesday, March 18, 2003

8:00 a.m. Breakfast

8:30 Roundtable Discussion:

I. Industry Approaches to Basic Research, Transition to Advanced Development, Advanced Development, and Licensure for Vaccines and Therapeutic Drugs

II. Challenges or Barriers to Additional Industry Involvement in Research and Development of Medical Countermeasures Against Biological Warfare Agents

Maria Bernwinkler, Director, Clinical Research, PPD
LTC Robert Borowski, Ph.D., Senior Medical Advisor, Office of the Deputy to the Assistant to the Secretary of Defense for Chemical and Biological Defense, DoD
COL David Danley, Ph.D., Director, Chemical and Biological Medical Systems, DoD
Joan Fusco, Ph.D., Vice President, Technical Affairs, Baxter Vaccines
Karen Goldenthal, M.D., Director, Division of Vaccine and Related Products Applications, Center for Biologics Evaluation and Research, FDA
Christine Grant, J.D., M.B.A., Vice President, Public Policy and Government Relations, Aventis Pasteur
Carole Heilman, Ph.D., Director, Division of Microbiology and Infectious Diseases, NIAID, NIH
COL Charles Hoke, Jr., M.D., U.S. Army Medical Research and Materiel Command (USAMRMC), DoD
Raymond Keifer, Ph.D., Senior Director Manufacturing, BioReliance Corporation
Michael Langford, D.V.M., Ph.D., Senior Vice President and Chief Scientific Officer, DynPort Vaccine Company, LLC
Frank Lee, Ph.D., President and CEO, engeneOS
Brad Leissa, M.D., Office of Counterterrorism and Pediatric Drug Development, Center for Drug Evaluation and Research, FDA
John McCormick, M.D., Orphan Products Development, FDA
Dennis Panicali, Ph.D., Chief Scientific Officer, Therion Biologics
Jerald Sadoff, M.D., Washington, DC
Peter Young, M.B.A., President and CEO, AlphaVax
Philip Youngman, Ph.D., Vice President of Discovery Biology, Elitra Pharmaceuticals

11:45 Adjourn Open Session

MEETING IV
APRIL 22, 2003

The National Academy of Sciences Building
2101 Constitution Avenue, N.W.
Washington, DC

Tuesday, April 22, 2003

9:00 a.m. Breakfast

9:30 Introductory Remarks
Leslie Z. Benet, Ph.D., Chair, Committee on Accelerating the Research, Development, and Acquisition of Medical Countermeasures Against Biological Warfare Agents

9:45 Availability of Biocontainment Facilities Needed to Support Research and Development of Medical Countermeasures to Biological Warfare Agents
Gerald Parker, M.S., D.V.M., Ph.D., Colonel, Veterinary Corps, Assistant Deputy for Research and Development, USAMRMC

Discussion

11:15 Break

11:30 Role of the Assistant Secretary of Defense for Health Affairs in DoD Efforts Toward Accelerating the Development and Licensure of Medical Countermeasures Against Biological Warfare Agents
William Winkenwerder, Jr., M.D., M.B.A., Assistant Secretary of Defense for Health Affairs

Discussion

12:30 p.m. Working Lunch

1:15 The Role That New Technologies (Genomics, Proteomics) Might Play in the Acceleration of Drug and Vaccine Research and Development
Stephen Hoffman, M.D., Sanaria

1:45	Rapid Development of a Medical Countermeasure to Protect Against Anthrax *Vivian Albert, Ph.D., Human Genome Sciences*
2:15	Discussion
3:00	Adjourn Open Session

MEETING V
JUNE 2, 2003

The National Academies Keck Center
500 Fifth Street, N.W.
Washington, DC

Monday, June 2, 2003

1:45 p.m.	Introductory Remarks *Leslie Z. Benet, Ph.D., Chair, Committee on Accelerating the Research, Development, and Acquisition of Medical Countermeasures Against Biological Warfare Agents*
2:00	Perspectives on Liability Issues Related to Biowarfare Countermeasures *Mr. Alan B. Morrison, LL.B., Public Citizen Litigation Group* Discussion
3:00	Intellectual Property Concerns of the Pharmaceutical Industry *Mr. Jeffrey Kushan, J.D., Sidley Austin Brown and Wood* Discussion, Including Government Contracts Perspective *Mr. Jay Winchester (by phone), U.S. Army Medical Research and Materiel Command*
4:30	Attributes of the Lieberman–Hatch Proposed Legislation *Mr. Chuck Ludlam, J.D., Office of Senator Joseph Lieberman*
5:30	Adjourn Open Session

APPENDIX C

Acknowledgments

The committee is grateful to the following individuals who provided information and assistance to the study through presentations or participation at meetings and workshops, informal meetings, or other means. Affiliations listed are those at the time of the individual's first contact with the committee.

Vivian Albert, Ph.D., Vice President, Preclinical Research and Development, Human Genome Sciences

Penrose (Parney) Albright, Ph.D., Senior Advisor, Science and Technology Directorate, Department of Homeland Security

Maria Bernwinkler, Director, Clinical Research, PPD

LTC Robert Borowski, Ph.D., Senior Medical Advisor, Office of the Deputy Assistant to the Secretary of Defense for Chemical and Biological Defense

COL W. Neal Burnette, Ph.D., Assistant Joint Program Executive Officer, Medical Systems, Joint Program Executive Office for Chemical and Biological Defense Programs

John Carney, Ph.D., Unconventional Pathogen Countermeasures Program, Defense Advanced Research Projects Agency

Salvatore Cirone, D.V.M., M.P.V.M., Program Director, Health Science Policy, Force Health Protection and Readiness, Office of the Assistant Secretary of Defense (Health Affairs)

LTC Edward Clayson, Ph.D., Chemical Biological Medical Systems, Joint Program Executive Office for Chemical and Biological Defense

COL David Danley, Ph.D., Director, Chemical Biological Medical Systems, Joint Program Executive Office for Chemical and Biological Defense

David Edman, Ph.D., Chemical Biological Medical Systems, Joint Program Executive Office for Chemical and Biological Defense

William Egan, Ph.D., Deputy Director, Office of Vaccines Research and Review, Food and Drug Administration

Anthony S. Fauci, M.D., Director, National Institute of Allergy and Infectious Diseases

Richard Fieldhouse, Professional Staff Member, Senate Armed Services Committee

MAJ Ronald Fizer, Joint Requirements Integrator, Joint Requirements Office for Chemical, Biological, Radiological, and Nuclear Defense

Joan Fusco, Ph.D., Vice President, Technical Affairs, Baxter Vaccines

Karen Goldenthal, M.D., Director, Division of Vaccines and Related Products Applications, Center for Biologics Evaluation and Research, Food and Drug Administration

Jesse Goodman, M.D., M.P.H., Director, Center for Biologics Evaluation and Research, Food and Drug Administration

Christine Grant, J.D., M.B.A., Vice President, Public Policy and Government Relations, Aventis Pasteur

Carole Heilman, Ph.D., Director, Division of Microbiology and Infectious Diseases, National Institute of Allergy and Infectious Diseases

COL Erik Henchal, Ph.D., Commander, U.S. Army Medical Research Institute of Infectious Diseases

Stephen Hoffman, M.D., Chief Executive and Scientific Officer, Sanaria Inc.

COL Charles Hoke, Jr., M.D., U.S. Army Medical Research and Materiel Command

Terry Irgens, President, DynPort Vaccine Company LLC

Leonard Izzo, Technical Director, Joint Requirements Office for Chemical, Biological, Radiological, and Nuclear Defense

Virginia Johnson, Ph.D., Vice President for Regulatory Affairs, DynPort Vaccine Company LLC

Anna Johnson-Winegar, Ph.D., Deputy Assistant to the Secretary of Defense for Chemical and Biological Defense

Raymond Keifer, Ph.D., Senior Director Manufacturing, BioReliance Corporation

Lawrence Kerr, Ph.D., Assistant Director for Homeland Security, Office of Science and Technology Policy

Dale Klein, Ph.D., Assistant to the Secretary of Defense for Nuclear and Chemical and Biological Defense

Rudy Kuppers, Ph.D., SAIC Support Contractor, Military Infectious Diseases Research Program, U. S. Army Medical Research and Materiel Command
Jeffrey Kushan, J.D., Partner, Sidley Austin Brown and Wood
Michael J. Langford, D.V.M., Ph.D., Senior Vice President and Chief Scientific Officer, DynPort Vaccine Company LLC
Frank Lee, Ph.D., President and CEO, engeneOS
Brad Leissa, M.D., Deputy Director, Division of Counterterrorism, Center for Drug Evaluation and Research, Food and Drug Administration
Stanley Lemon, M.D., Dean of Medicine and Professor, University of Texas Medical Branch, Galveston
Carol Linden, Ph.D., Director, Medical Chemical and Biological Defense Research Program, U.S. Army Medical Research and Materiel Command
Chuck Ludlam, J.D., Counsel, Office of Senator Joseph Lieberman
MG Lester Martinez-Lopez, M.D., Commander, U.S. Army Medical Research and Materiel Command
Mark B. McClellan, M.D., Ph.D., Commissioner, Food and Drug Administration
John McCormick, M.D., Orphan Products Development, Food and Drug Administration
Andrea Meyerhoff, M.D., Director, Counterterrorism Programs, Office of the Commissioner, Food and Drug Administration
Alan B. Morrison, LL.B., Public Citizen Litigation Group
Dianne Murphy, M.D., Director, Office of Counterterrorism and Pediatric Drug Development, Center for Drug Evaluation and Research, Food and Drug Administration
Col Joseph Palma, M.D., M.P.H., Chief Medical Advisor, Office of the Deputy Assistant to the Secretary of Defense for Chemical and Biological Defense
Dennis Panicali, Ph.D., Chief Scientific Officer, Therion Biologics
COL Gerald Parker, M.S., D.V.M., Ph.D., Assistant Deputy for Research and Development, U.S. Army Medical Research and Materiel Command
Vicki Pierson, Ph.D., Biodefense Projects Manager, Office of Biodefense Research Affairs, National Institute of Allergy and Infectious Diseases
Rick Prouty, Medical Action Officer, Joint Requirements Office for Chemical, Biological, Radiological, and Nuclear Defense
Gerald V. Quinnan, Jr., M.D., Professor and Chair of Preventive Medicine, Uniformed Services University of the Health Sciences

Jean Reed, Professional Staff Member, House Armed Services Committee
BG Stephen Reeves, Program Executive Officer, Chemical and Biological Defense
Philip Russell, M.D., Office of Emergency Preparedness Planning, Department of Health and Human Services
Jerald Sadoff, M.D., Washington, DC
Lynn Selfridge, Chemical Biological Medical Systems, Joint Program Executive Office for Chemical and Biological Defense
Leonard Smith, Ph.D., Department of Immunology and Molecular Biology, Division of Toxicology, U.S. Army Medical Research Institute of Infectious Diseases
John D. Strandberg, D.V.M., Ph.D., Director, Division of Comparative Medicine, National Center for Research Resources, National Institutes of Health
COL David Vaughn, M.D., M.P.H., Research Area Director, Military Infectious Diseases Research Program, U.S. Army Medical Research and Materiel Command
LTC Keith R. Vesely, D.V.M., Ph.D., Chief, Medical Science and Technology Division, Chemical and Biological Defense Directorate, Defense Threat Reduction Agency
Judith Wagner, Ph.D., Scholar in Residence, Institute of Medicine
Jay Winchester, Office of the Staff Judge Advocate, U.S. Army Medical Research and Materiel Command
William Winkenwerder, Jr., M.D., M.B.A., Assistant Secretary of Defense for Health Affairs
Janet Woodcock, M.D., Director, Center for Drug Evaluation and Research, Food and Drug Administration
Peter Young, M.B.A., President and CEO, AlphaVax
Philip Youngman, Ph.D., Vice President of Discovery Biology, Elitra Pharmaceuticals
Kathryn Zoon, Ph.D., Director, Center for Biologics Evaluation and Research, Food and Drug Administration

APPENDIX D

Biographical Sketches

Leslie Z. Benet, Ph.D. (*Chair*), is a professor and former chairman of the Department of Biopharmaceutical Sciences at the University of California, San Francisco. His research interests, more than 440 publications, and 10 patents are in the areas of pharmacokinetics, biopharmaceutics, and pharmacodynamics. His most recent work has addressed the interplay of metabolic enzymes and transport proteins as related to the disposition of immunosuppressive, anticancer, anti-AIDS, and antiparasitic drugs, as well as drugs of importance to women's health. He is a fellow of the American Association for the Advancement of Science, the American Association of Pharmaceutical Scientists (AAPS), and the Academy of Pharmaceutical Research and Science. He is the chairman of the board at AvMax, Inc., and OxoN Medica, Inc., and serves as a consultant to several pharmaceutical and biotechnology companies. Dr. Benet is a recipient of the AAPS Distinguished Pharmaceutical Scientist Award, the American Pharmaceutical Association Higuchi Research Prize, the American Society for Clinical Pharmacology Rawls-Palmer Award for Progress in Medicine, the International Pharmaceutical Federation (FIP) Høst-Madsen Medal, and five honorary doctorates. He previously served as chair for the Food and Drug Administration (FDA) Center for Biologics Evaluation and Research External Peer Review Committee, the FDA Expert Panel on Individual Bioequivalence, and the FIP Board of Pharmaceutical Sciences and as a member of the FDA Science Board and the Board of Directors of the American Foundation for Pharmaceutical Education. An elected member of the Institute of Medicine (IOM), Dr. Benet has served as the chair or a member of various IOM committees; he is currently a member of the Board on Health Sciences Policy.

Walter E. Brandt, Ph.D., is a senior program officer for the Malaria Vaccine Initiative (MVI) at the Program for Appropriate Technology in Health (PATH), an organization focused on the acceleration of the development of malaria vaccine candidates. He also serves on the scientific advisory council for the Sabin Vaccine Institute. He previously served as chair for the World Health Organization Subcommittee on Dengue and Japanese Encephalitis Vaccines and as chair of a National Vaccine Program Office subcommittee on the safety of a vaccinia-vectored rabies vaccine. Before joining MVI, Dr. Brandt was a senior scientist at Science Applications International Corporation where he advised on vaccine development strategy and plans and prepared documents for the Army on vaccines for anthrax, botulism, and plague for FDA review. As a project manager in the U.S. Army Medical Research and Materiel Command he had oversight of the development, manufacture, and testing of vaccines and immune globulins, responsibility for more than 50 active Investigational New Drug applications with the FDA, and management of resources and activities that contributed to licensure of four vaccines. As a microbiologist and assistant chief of the Department of Virus Diseases, Walter Reed Army Institute of Research, he worked on teams studying dengue fever virus infections and the immune response, markers for live attenuated vaccine candidates, and characterization of reagents for diagnostic tests.

Barry S. Coller, M.D., serves as the David Rockefeller Professor of Medicine, head of the laboratory of blood and vascular biology, and vice president for medical affairs at the Rockefeller University. Dr. Coller's research is devoted primarily to investigating platelet physiology, vascular biology, and adhesion phenomena in sickle cell disease. He is responsible for the production of the monoclonal antibody that later was modified to become the drug abciximab (ReoPro), which is used to prevent ischemic complications of percutaneous coronary interventions and unstable angina. He has served on professional society committees and held several professional society offices and editorial positions, including coediting the fifth and sixth editions of *Williams Hematology*. He is on the Board of Governors of the National Institutes of Health (NIH) Clinical Center and on the Board of Extramural Advisors of the National Heart, Lung, and Blood Institute. He has served as a consultant to Centocor, Inc., and Genentech. Dr. Coller was elected a member of the IOM in 1999 and the National Academy of Sciences in 2003. He is a scientific advisor to two biotechnology companies.

Glenna M. Crooks, Ph.D., is the founder and president of Strategic Health Policy International, Inc. (SHPI), an organization devoted to assisting businesses and governments with the management of health care policy and political issues. Dr. Crooks's clients include pharmaceutical and vaccine companies. She gained experience in public policy, health care, and line management in government and business during her tenure as vice president for worldwide operations of Merck's vaccine business and as Deputy Assistant Secretary for Health in the Reagan administration. Dr. Crooks serves on the board of directors for Partnership for Prevention and was formerly on the board of American Biogenic Sciences, Inc., chairman of the National Commission on Rare Diseases, and a member of the National Council of the Institute for Child Health and Human Development. Her awards include the Surgeon General's Medallion, awarded by C. Everett Koop, and the Congressional Exemplary Service Award for Orphan Products Development.

R. Gordon Douglas, Jr., M.D., is a consultant in infectious diseases, vaccines, and global health. In 1999 he retired from Merck after serving as president of the Vaccine Division and a member of the Management Committee for nine years. He is currently director of strategic planning of the Vaccine Research Center, National Institute of Allergy and Infectious Diseases and adjunct professor of medicine at Cornell University. He also serves as a director of the biotechnology companies Vical, Elusys Therapeutics, Advancis Pharmaceutical Corporation, Iomai, and VaxInnate and the nonprofit organizations International AIDS Vaccine Initiative and Aeras Global TB Vaccine Foundation. Dr. Douglas is a graduate of Princeton University and Cornell University Medical College. He served as head of the Infectious Diseases Clinic at the University of Rochester (1970–1982) and chairman of the Department of Medicine at New York Hospital, Cornell University (1982–1990) before joining Merck. He is author of more than 200 original scientific publications dealing with viral pathogenesis, vaccines, and antivirals, and was co-editor of *Principals and Practices of Infectious Diseases*, the standard reference in the field. Dr. Douglas is the recipient of the R.R. Hawkins Award (Association of American Publishers, 1980), the Harry Feldman award (Infectious Diseases Society of America, 1992) and the Maxwell Finland Award (National Foundation for Infectious Diseases, 2000). He is also a member of IOM, the Association of American Physicians, and the American Society of Clinical Investigators. Dr. Douglas served on the IOM committee that issued the report *Protecting Our Forces: Improving Vaccine Acquisition and Availability in the U.S. Military*.

Jacques S. Gansler, Ph.D., is the Roger C. Lipitz Chair in Public Policy and Private Enterprise at the University of Maryland School of Public Affairs. He served as the Under Secretary of Defense for Acquisition, Technology, and Logistics from 1997 to 2001. As the third-ranking civilian in the Department of Defense, Dr. Gansler was responsible for acquisition, research and development, logistics, advanced technology, international programs, environmental security, and nuclear, chemical, and biological programs. Prior to his appointment, he served in several other roles in the private technology sector and as Deputy Assistant Secretary of Defense (Materiel Acquisition). He has published and taught on subjects related to the defense industry and served as an honorary professor at the Industrial College of the Armed Forces as well as a visiting professor at the University of Virginia. Dr. Gansler is a member of the National Academy of Engineering.

Anthony L. Itteilag is an independent consultant on business and management issues pertaining to the federal government. Prior to consulting, he served in multiple capacities at the NIH: senior advisor to the director, deputy director for management and chief financial officer, as well as acting chief information officer. He continues to work as a volunteer for the Office of the Director at NIH. He has employed his budget and program analysis experience at the Departments of Health and Human Services (formerly Health, Education and Welfare), Interior, and Defense for almost four decades. He is a member of the American Society for Public Administration and a graduate of the Federal Executive Institute. Over the course of his career, Mr. Itteilag has received numerous public service awards, including the Clifford R. Gross Award for Federal Public Service from the American Society for Public Administration (Maryland Chapter) in 2001, four Senior Executive Service presidential rank awards, Distinguished Service Awards from the Secretaries of the Interior and Health and Human Services, and an exemplary service award from the Surgeon General.

Dennis L. Kasper, M.D., is executive dean for academic programs, William Ellery Channing Professor of Medicine, and professor of microbiology and molecular genetics at Harvard Medical School. He also serves as director of the Channing Laboratory and as a senior physician at Brigham and Women's Hospital. With his colleagues and students, Dr. Kasper studies the molecular basis of bacterial pathogenesis, applying the resulting knowledge to enhance understanding of the interactions of bacterial surface virulence factors with host defenses. Dr. Kasper's studies focus on the molecular and chemical characterization of important bacterial virulence factors such as capsular polysaccharides, surface proteins, and toxins. The ultimate goal is to develop vaccines and immunomodulatory molecules to prevent bacterial infections and their complications. Dr. Kasper is a fellow of the American Academy of Microbiology and the American Association for the Advancement of Science as well as a member of IOM. He is a consultant to two pharmaceutical companies and serves on the scientific advisory board of Microbia.

Steven Kelman, Ph.D., is Albert J. Weatherhead III and Richard W. Weatherhead Professor of Public Management at the John F. Kennedy School of Government, Harvard University, where he conducts research on public sector operations management with a focus on organizational design and change. He is the former Office of Federal Procurement Policy administrator in the Office of Management and Budget, where he was involved in the Federal Acquisition Streamlining Act of 1994 and the Federal Acquisition Reform Act of 1995. He is the author of several books in the area of public policy, including his 1990 book entitled *Procurement and Public Management: The Fear of Discretion and the Quality of Government Performance*. He is currently completing research on the spread of procurement reform innovations at the working levels of government organizations. Dr. Kelman is on the editorial board of the *Journal of Public Administration Research and Theory* and is a fellow of the National Academy of Public Administration.

Richard F. Kingham, J.D., is a partner in the law firm of Covington & Burling, assigned to the Washington, DC, and London offices. Since joining the firm in 1973, he has concentrated on pharmaceutical regulation, product liability, and related issues. He represented vaccine manufacturers in connection with the swine flu program in 1976, the childhood vaccine injury compensation legislation in 1986, and legislative and administrative matters relating to vaccines and other products to protect against bioterrorism. He has served as a member of the National Advisory Allergy and Infectious Diseases Council of NIH and as a member of or advisor to committees of IOM. From 1977 to 1990, he served as a lecturer at the University of Virginia School of Law; he presently serves as an adjunct professor at the Georgetown University Law Center and lectures in the graduate program of pharmaceutical medicine of the University of Wales. He received his law degree from the University of Virginia in 1973.

Peter M. Palese, Ph.D., is a professor of microbiology and chairman of the Department of Microbiology at the Mount Sinai School of Medicine. He has more than 200 scientific publications that include research on the replication of RNA-containing viruses with a special emphasis on influenza viruses, which are negative-strand RNA viruses. He elucidated the genetic maps of influenza A, B, and C viruses and obtained precise measurements of their mutation rates, and also developed the reverse genetics system that for the first time enabled the manipulation and analysis of influenza and other minus-strand RNA viruses. He serves on the editorial board for the *Proceedings of the National Academy of Sciences* and as an editor for the *Journal of Virology*. He is a member of the Vaccines and Related Biological Products Advisory Committee, Center for Biologics Evaluation and Research at the FDA. Dr. Palese was elected to the National Academy of Sciences in 2000 for his seminal studies on influenza viruses.

Paul Parkman, M.D., retired as director of the Center for Biologics Evaluation and Research of the FDA in 1990, after 30 years of federal service (18 at FDA) in infectious disease research and biological product regulation. He is currently a consultant in the development and production of biologics, including both the traditional vaccines and the newer products of modern biotechnology. His research led to the discovery of rubella virus and diagnostic tests for rubella. He and his colleagues developed and tested the first successful experimental rubella vaccine in 1965. The use of rubella vaccines, starting in 1969, produced major reductions of this disease and the consequent congenital defects common in babies whose mothers had been infected early in pregnancy.

Ronald J. Saldarini, Ph.D., is currently a consultant to several biotechnology companies and is a director at Alphavax, Medarex, and Cellegy Pharmaceuticals. He is chairman of Therion Biologics and Idun Pharmaceuticals. He is an associate in Naimark and Associates, consultants to the healthcare industry, and is also a consultant to the Malaria Vaccine Initiative. From 1986 until his retirement in 1999, he was the president of the global vaccine business of American Cyanamid (Lederle Praxis) and American Home Products (Wyeth Lederle). He has been a member of the Board of Trustees of the National Foundation of Infectious Diseases, the Infectious Disease Institute of New Jersey, the Immunization Advisory Council of the New York State Department of Health, the Corporate Council for the Children's Health Fund, the policy board of the Albert B. Sabin Vaccine Foundation, and the Board of Directors of the Institute for Advanced Studies of Immunology and Aging. Recently, Dr. Saldarini was a member of the IOM committee that issued the report *Protecting Our Forces: Improving Vaccine Acquisition and Availability in the U.S. Military*; he also served on the IOM Committee on Immunization Finance Workshops.

Jane E. Sisk, Ph.D., is an economist and professor in the Department of Health Policy and a co-director of the Center on Evidence-Based Medicine and Aging at the Mount Sinai School of Medicine. Her current research areas are the evaluation of strategies to improve quality of care with a focus on the reduction of disparities among population subgroups; the organizational and financial arrangements for the delivery of care, including managed care and Medicaid; and the cost-effectiveness of health care interventions, with an emphasis on preventive care. Prior to joining the faculty at Mount Sinai, Dr. Sisk was a professor of health policy at Columbia University. Before that, she was a senior associate and project director on health policy projects at the congressional Office of Technology Assessment. She has served on several IOM committees over the last 20 years and was elected to IOM in 2001. She currently serves as a member on the IOM Committee on Medicare Coverage of Routine Thyroid Screening. In addition to her involvement with IOM, Dr. Sisk has been a member of the Study Section on Health Care Quality and Effectiveness for the U.S. Agency for Healthcare Research and Quality.

Elaine I. Tuomanen, M.D., is director of the Children's Infection Defense Center and chair of the Department of Infectious Diseases at St. Jude Children's Research Hospital in Memphis. She is a board-certified pediatrician with subspecialty training in pediatric infectious diseases. Previously, Dr. Tuomanen was head of the Laboratory of Molecular Infectious Diseases at the Rockefeller University. Her research focus is the induction of inflammation and the molecular pathogenesis of infection by gram-positive *Streptococcus pneumoniae* bacteria, with a view to understanding the host–pathogen relationship. She has a strong interest in the development of novel therapeutics, including vaccines and antibiotics, and is currently involved in the development of a new Good Manufacturing Practice (GMP) manufacturing facility at St. Jude's.

Benjamin J. Weigler, D.V.M., M.P.H., Ph.D., is associate professor of comparative medicine in the School of Medicine and adjunct associate professor of epidemiology in the School of Public Health at the University of Washington, Seattle. He also serves as director of animal health resources for the Fred Hutchinson Cancer Research Center in Seattle. He is board-certified by the American College of Laboratory Animal Medicine and the American College of Veterinary Preventive Medicine. His research is in the area of infectious disease epidemiology and public health, with concentration in zoonotic diseases. Dr. Weigler is a member of several professional veterinary medical organizations and is on the editorial board for the scientific journal *Comparative Medicine*. He recently served on the National Research Council (NRC) Committee on Occupational Health and Safety in Care of Nonhuman Primates and the steering committee for the

Centers for Disease Control and Prevention and Eagleson Institute Seventh National Biosafety Symposium in Atlanta.

Janet Westpheling, Ph.D., is an associate professor of genetics at the University of Georgia. Her primary areas of research involve the control of gene expression in *Streptomyces* bacteria, with emphasis on the study of carbon utilization and primary metabolism, as well as the strategies used by bacteria to regulate genes involved in morphogenesis and antibiotic production. *Streptomyces* are of particular interest because they produce most of the natural product antibiotics used in human and animal health care. Dr. Westpheling serves on the *Journal of Bacteriology* editorial board and was chair of the Gordon Research Conference on Microbial Stress Response in 1996. She serves as a member of the scientific advisory boards of several biotechnology companies interested in natural product drug discovery and is a consultant to pharmaceutical and biotechnology companies. Dr. Westpheling participates annually in a course on fermentation technology offered by the Chemical Engineering Department at the Massachusetts Institute of Technology. She has previously served on three NRC committees: the Committee on Opportunities in Biotechnology for Future Army Applications, the Committee on Biobased Industrial Products: National Research and Commercialization Priorities, and the Committee on Bioprocess Engineering.

Board of the Medical Follow-up Agency Liaison to the Committee

Timothy R. Gerrity, Ph.D., is the founding director of the Bioengineering Institute at Worcester Polytechnic Institute (WPI) and a research professor in the Department of Biomedical Engineering. The Bioengineering Institute is dedicated to translational research converting the products of research into usable medical devices. Before joining WPI, Dr. Gerrity was executive director of the Georgetown School of Medicine Chronic Pain and Fatigue Research Center, which focuses on a multidisciplinary approach to understanding the underlying mechanisms of chronic multisymptom illness. He has also served as special assistant chief research and development officer (1997–1999) and deputy director for medical research (1994–1997) at the Department of Veterans Affairs (VA). Earlier, he carried out air pollution research with the Environmental Protection Agency. Dr. Gerrity's research interests have included the effects of air pollution on human cardiopulmonary function, improved understanding of the behavior of inhaled aerosols in the human lungs, and the mechanisms by which particles are cleared from the lungs. Dr. Gerrity received his B.S., M.S., and Ph.D. degrees in physics from the University of Illinois at Chicago and postdoctoral training in pulmonary physiology from the Department of Medicine at the University of Illinois.

Consultant to the Committee

James D. Marks, M.D., Ph.D., is professor of anesthesia and pharmaceutical chemistry at the University of California, San Francisco. He is board certified in internal medicine, anesthesia, and critical care medicine. From 1996 to 2001 he was the medical director of the Medical-Surgical Intensive Care Unit at San Francisco General Hospital and continues to attend in the intensive care unit and operating rooms there. Dr. Marks is a pioneer in the field of antibody engineering, where he has developed widely used technology for generating and optimizing human therapeutic antibodies. He currently directs a research group using antibody gene diversity libraries and display technologies to dissect the molecular basis of infectious diseases and cancer and to develop novel antibody-based therapeutic approaches for these diseases. In the field of oncology, his research has elucidated the impact of antibody biophysical properties on tumor targeting, and his laboratory has generated a novel antibody-based drug that is being commercialized for breast cancer therapy. His laboratory is also funded by the Department of Defense and the National Institute of Allergy and Infectious Diseases (NIAID) to develop antibody-based therapies for the biothreat agent botulinum neurotoxin. Dr. Marks has served on Health and Human Services and NIAID expert advisory panels on the botulinum neurotoxins. He has more than 110 publications in the field of antibody engineering and is an inventor on 62 issued or pending patents.